D1710263

OT 6:
Operator Theory: Advances and Applications
Vol. 6

Edited by

I. Gohberg

Birkhäuser Verlag
Basel · Boston · Stuttgart

Invariant Subspaces and Other Topics

6th International Conference on Operator Theory,
Timişoara and Herculane (Romania),
June 1–11, 1981

Volume Editors

C. Apostol
R. G. Douglas
B. Sz.-Nagy
D. Voiculescu

Managing Editor

Gr. Arsene

1982

Birkhäuser Verlag
Basel · Boston · Stuttgart

Volume Editorial Office

Department of Mathematics
INCREST
Bd. Păcii 220
79622 Bucharest (Romania)

CIP-Kurztitelaufnahme der Deutschen Bibliothek

Invariant subspaces and other topics / 6th Internat. Conference on Operator Theory, Timişoara and Herculane (Romania), June 1–11, 1981. Vol. ed. C. Apostol ... Managing ed. Gr. Arsene. – Basel ; Boston ; Stuttgart : Birkhäuser, 1982.
(Operator theory ; Vol. 6)
ISBN 3-7643-1360-9
NE: Apostol, Constantin [Hrsg.]; International Conference on Operator Theory ‹06, 1981, Timişoara; Herculane›; GT

Printed in Switzerland by Birkhäuser AG, Graphisches Unternehmen, Basel
ISBN 3-7643-1360-9

CONTENTS

Preface 7

List of participants 9

Program of the conference 12

AKEMANN, C.A.; ANDERSON, J. — The Stone-Weierstrass problem for C^*-algebras 15

APOSTOL, C. — On the norm-closure of the similarity orbit of essentially nilpotent operators 33

BEAUZAMY, B. — Invariant subspaces and functional representations for the $C_{\cdot 1}$ contractions 45

CHEVREAU, B. — Intertwining and hyperinvariant subspaces 51

COWEN, M.J.; DOUGLAS, R.G. — On moduli for invariant subspaces................................ 65

GOLOGAN, R.-N. — An extension of Chacon-Ornstein ergodic theorem......................... 75

JANAS, J. — Commuting subnormal operators quasimilar to multiplication by coordinate functions on odd spheres 81

JONES, V.; POPA, S. — Some properties of MASA's in factors................................ 89

LANGER, H.; TEXTORIUS, B. — Generalized resolvents of dual pairs of contractions 103

LARSON, D.R. — Annihilators of operator algebras 119

LU, Shijie — On the derivations with norm closed range in Banach spaces........... 131

NAGY, B. — On boolean algebras of projectors and prespectral operators 145

NIKOLSKII, N.K.; VASJUNIN, V.I. — Control subspaces of minimal dimension, and spectral multiplicity 163

PELIGRAD, C. — Derivations of C^*-algebras which are invariant under an automorphism group. II......................... 181

POPA, S. — On commutators in properly infinite W^*-algebras 195

SUCIU, I. — A functional model for the unitary dilation of a positive definite map 209

TIMOTIN, D. — The Levinson algorithm in linear prediction 217

ZEMÁNEK, J. — Geometric interpretation of the essential minimum modulus 225

PREFACE

The annual Operator Theory conferences in Timişoara are conceived as a means to promote cooperation and exchange of information between specialists in all areas of Operator Theory. The present volume consist of papers contributed by the participants of the 1981 Conference. Since many of these papers contain results on the invariant subspace problem or are related to the role of invariant subspaces in the study of operators or operator systems, we thought it appropiate to mention this in the title of the volume, though the "other topics" have a wide range. As in past years, special sessions concerning other fields of Functional Analysis were organized at the 1981 Conference, but contributions to these sessions are not included in the present volume.

The research contracts of the Department of Mathematics of INCREST with the National Council for Sciences and Technology of Romania provided the means for developping the research activity in Functional Analysis; these contracts constitute the generous framework for these meetings.

We want also to acknowledge the support of INCREST and the excelent organizing job done by our host - University of Timişoara-. Professor Dumitru Gaşpar and Professor Mircea Reghiş are among those people in Timişoara who contributed in an essential way to the success of the meeting.

We are indebted to Professor Israel Gohberg for including these Proceedings in the OT Series and for valuable help in the editing process. Birkhäuser Verlag was very cooperative in publishing this volume.

Rodica Gervescu and Camelia Minculescu dealt with the difficult task of typing the whole manuscript; they did an excelent job in a very short time.

Organizing Committee

Head of Math.Department of INCREST,

Zoia Ceauşescu

Organizers,

Constantin Apostol

Dan Voiculescu

LIST OF PARTICIPANTS*)

ALBRECHT, Ernst	University of Saarlandes, West Germany
ANDERSON, Joel	Pennsylvania State University, USA
ANGHELINA, Elisabeta	University of Timişoara
APOSTOL, Constantin	INCREST, Bucharest
ARSENE, Grigore	INCREST, Bucharest
BALINT, Ştefan	University of Timişoara
BÂNZARU, Titus	Politechnical Institute, Timişoara
BEAUZAMY, Bernard	Claude Bernard University, Lyon, France
BERGER, Charles A.	Yeshiva University, New York, USA
BIRĂUŞ, Silviu	University of Timişoara
BOGNĂR, Janos	Mathematics Institute, Budapest, Hungary
BOJA, Nicolae	Politechnical Institute, Timişoara
BUCUR, Gheorghe	INCREST, Bucharest
CÂMPU, Eugen	University of Bucharest
CEAUŞESCU, Zoia	INCREST, Bucharest
CEAUŞU, Traian	Politechnical Institute, Timişoara
CHEVREAU, Bernard	University of Bordeaux, France
COBZAŞ, Ştefan	Babeş-Bolyai University, Cluj-Napoca
COLOJOARĂ, Ion	University of Bucharest
COLOJOARĂ, Sanda	Politechnical Institute, Bucharest
CONSTANTINESCU, Tiberiu	INCREST, Bucharest
CORDUNEANU, Adrian	Politechnical Institute, Iaşi
COSTINESCU, Roxana	University of Bucharest
CRAIOVEANU, Mircea	Politechnical Institute, Timişoara
CRSTICI, Boris	Politechnical Institute, Timişoara
D'ANTONI, Claudio	University of Roma, Italy
DEUTSCH, Emerich	Politechnical Institute of New York,USA
DINESCU, Gabriela	University of Bucharest
DOUGLAS, Ronald G.	State University of New York, USA
DRAGOMIR, Achim	University of Timişoara
ECKSTEIN, Gheorghe	University of Timişoara
FAOUR, Nazih	University of Kuwait
FRUNZĂ, Ştefan	University of Iaşi
GAŞPAR, Dumitru	University of Timişoara
GĂVRUŢĂ, Pascu	Politechnical Institute, Timişoara
GHEONDEA, Aurelian	INCREST, Bucharest
GIURGIU, Maria	Military Academy, Bucharest
GODINI, Gliceria	INCREST, Bucharest
GOLOGAN,Radu	INCREST, Bucharest
GONG, Weibang	Qufu Teachers College, Shangton, China
HĂRĂGUŞ, Dumitru	University of Timişoara
HATVANY, Csaba	Politechnical Institute, Timişoara

*) Romanian participants are listed only with the name of their institution.

HIRIŞ, Viorel	University of Timişoara
HO SI HAU	University of Hanoi, Vietnam
JANAS, Jan	Mathematics Institute, Krakow, Poland
KÉRCHY, Lajos	University of Szeged, Hungary
LANGER, Heinz	Technical University of Dresden, GDR
LARSON, David R.	University of Nebraska, Lincoln, USA
LIPOVAN, Octavian	Politechnical Institute, Timişoara
LU, Shijie	University of Nankin, China
MARTIN, Mircea	INCREST, Bucharest
MĂRUŞTER, Ştefan	University of Timişoara
MEGAN, Mihail	University of Timişoara
MÜLLER, Vladimir	Mathematics Institute, Prague, Czechoslovakia
MUSTAŢĂ, Constantin	Babeş-Bolyai University, Cluj-Napoca
MUSTAŢĂ, Paul	University of Bucharest
NAGY, Béla	Budapest Technological University,Hungary
NEAGU, Mihai	Politechnical Institute, Timişoara
NÉMETH, Alexandru	Babeş-Bolyai University, Cluj-Napoca
NICULESCU, Constantin	University of Craiova
NIKOLSKII, Nikolai K.	Steklov Institute, Leningrad, USSR
NOAGHI, Sorin	University of Timişoara
OCNEANU, Adrian	INCREST, Bucharest
PASCU, Mihai	University of Bucharest
PASNICU, Cornel	INCREST, Bucharest
PĂUNESCU, Doru	University of Timişoara
PELIGRAD, Costel	INCREST, Bucharest
PETZ, Denes	Mathematics Institute,Budapest, Hungary
PIMSNER, Mihai	INCREST, Bucharest
POPA, Constantin	University of Timişoara
POPA, Nicolae	INCREST, Bucharest
POPA, Sorin	INCREST, Bucharest
POPESCU, Nicolae	University of Timişoara
POTRA,Florian A.	INCREST, Bucharest
POTRA, Teodor	Babeş-Bolyai University, Cluj-Napoca
PTÁK, Vlastimil	Mathematics Institute,Prague,Czechoslovakia
PUTA, Mihai	University of Timişoara
PUTINAR, Mihai	INCREST, Bucharest
REGHIŞ, Mircea	University of Timişoara
RUS, Ion	Babeş-Bolyai University, Cluj-Napoca
ŞABAC, Mihai	University of Bucharest
ŞERB, Ioan	Babeş-Bolyai University, Cluj-Napoca
ŞUCIU, Ion	INCREST, Bucharest
SUCIU, Nicolae	University of Timişoara
SULJAGIC, Salih	University of Zagreb, Jugoslavia
STAN, Ilie	Politechnical Institute, Timişoara
SZÖKEFALVI-NAGY,BÉLA	University of Szeged, Hungary
TALPEŞ, Florin	University of Bucharest
TELEMAN, Silviu	INCREST, Bucharest
TERESCENCO, Alexandru	Computer Center of Timişoara
TIBA, Dan	INCREST, Bucharest
TIMOTIN, Dan	INCREST, Bucharest
TOPUZU, Paul	University of Timişoara
VALUŞESCU, Ilie	INCREST, Bucharest
VESCAN, Robert	Politechnical Institute, Iaşi
VOICULESCU, Dan	INCREST, Bucharest
VRABIE, Ioan	Politechnical Institute, Iaşi

YOUNG, Nicholas	University of Glasgow, Scotland
ZAHARIA, Dumitru	University of Timişoara
ZEMÁNEK, Jaroslav	Mathematics Institute, Prague,Czechoslovakia

PROGRAM OF THE CONFERENCE

TUESDAY, June 2

SECTION A

Chairman: R.G.Douglas

9:30-10:15 J.ANDERSON: Stone-Weierstrass theorems for separable C*-algebras.
10:30-11:15 C.PELIGRAD: On C*-dynamical systems.
11:30-12:15 B.NAGY: On prespectral operators.

SECTION A

Chairman: S.Popa

16:30-17:15 I.SUCIU: A functional model for the unitary dilation of a positive definite map.
17:25-17:55 V.MÜLLER: Non-removable ideals in a commutative Banach algebras.
18:05-18:35 R.GOLOGAN: Two ergodic theorems of Chacon-Ornstein type.
18:45-19:15 D.PETZ: Notes on reduction theory of von Neumann algebras.

SECTION B

Chairman: P.Mustaţă

16:00-16:30 N.POPESCU; M.REGHIŞ: L^2-controlability and stability for linear time varying systems in Banach spaces.
16:40-17:10 D.TIBA: Some existence results for partial differential equations.
17:20-17:50 E.DEUTSCH: The derivatives of the Perron roots of an irreducible nonnegative matrix.
18:00-18:30 M.PUTA: Spectral properties of the Laplace operator on differential forms.
18:40-19:10 S.COBZAŞ: The principle of condensation of singularities and applications.

WEDNESDAY, June 3

SECTION A

Chairman: C.Apostol

9:30-10:15 D.LARSON: Nest algebras and similarity transformations.
10:30-11:15 C.BERGER: A new class of dilation theorems.
11:30-12:15 B.BEAUZAMY: Invariant subspaces for $C_{\cdot 1}$ contractions.

SECTION A

Chairman: N.K.Nikolskii

16:30-17:15 S.Teleman: On the regularity of boundary measures.
17:25-17:55 N.BOBOC; GH.BUCUR: On archimedean measures.

SECTION B

Chairman: M.Reghiş

16:00-16:30 A.TERESCENCO: On the spectrum of Hardy kernels.
16:40-17:10 R.VESCAN: Quasi-variational inequalities solved by Ky Fan's Lemma.
17:20-17:50 I.VRABIE: A monotone convergence theorem in Banach spaces.
18:00-18:30 AL.NÉMETH: Positive linear operators with completely regular cone ranges.

THURSDAY, June 4

SECTION A

Chairman: H.Langer

9:30-10:15 E.ALBRECHT: Generalized multipliers.
10:30-11:00 M.PUTINAR; F.-H.VASILESCU: Continuous and analytic invariant for deformations of Fredholm complexes.

SECTION B

Chairman: D.Gaşpar

9:00-9:30 TH.POTRA: On some bases of spline functions in finite element method.
9:40-10:10 I.RUS: Coincidence and surjectivity.
10:20-10:50 A.DIACONU: On interpolation in abstract spaces.

FRIDAY, June 5

Chairman: J.Anderson

9:30-10:15 S.POPA: Maximal abelian *-subalgebras of von Neumann algebras.
10:30-11:15 A.OCNEANU: On the classification of discrete and compact actions on von Neumann algebras.
11:30-12:00 J.JANAS: Quasisimilar n-tuples of subnormal operators.

Chairman: C.Berger

16:30:17:00 N.YOUNG: Orbit structure of the unit sphere of $L(H)$ under the symplectic group.
17:10-17:40 L.KÉRCHY: On the commutant of C_{11}-contractions.
17:50-18:20 GR.ARSENE; FL.-A.POTRA: Extrapolation and prediction.

MONDAY, June 8

Chairman: D.Voiculescu

9:30-10:30 R.G.DOUGLAS: Index theory and K-homology.
11:00-12:00 B.SZ.-NAGY: Reflexive and hyperreflexive operators of class C_0.

Chairman: E.Albrecht

16:30-17:15 GR.ARSENE; A.GHEONDEA: Completing matrix contractions.
17:30-18:15 H.LANGER: Generalized resolvents and applications.
18:25-18:55 T.CONSTANTINESCU; A.GHEONDEA: Notes on selfadjoint extensions of symmetric operators.

TUESDAY, June 9

Chairman: B.Szökefalvi-Nagy

9:30-10:15 N.K.NIKOLSKII: Les systèmes linéaires controllables et la théorie de la multiplicité du spectre.
10:30-11:15 B.CHEVREAU: Intertwinings and hiperinvariant subspaces.
11:25-11:55 S.POPA: On commutators in properly infinite W*-algebras.
12:05-12:35 N.FAOUR: Toeplitz operators on Bergmann and Hardy spaces.

WEDNESDAY, June 10

Chairman: I.Suciu

9:30-10:15 J.ZEMÁNEK: A new look on the spectral radius formula.
10:30-11:15 LU SHIJIE : On the derivations with norm closed range in Banach spaces.
11:25-11:55 ŞT.FRUNZĂ: Spectral decomposition and duality for several operators.
12:05-12:35 D.TIMOTIN: Levinson algorithm in linear prediction.

THURSDAY, June 11

Chairman: V.Pták

9:30-10:15 C.APOSTOL: On the closure of the similarity orbit of essentially nilpotent operators.
10:25-10:55 C.D'ANTONI: Interpolation by type I factors and the flip automorphism.
11:05-11:35 G.DINESCU: Weak spectral equivalence.
11:45-12:15 P.GĂVRUŢĂ: Remarks on the weighted bilateral shift.

THE STONE-WEIERSTRASS PROBLEM FOR C^*-ALGEBRAS

Charles A.Akemann and Joel Anderson

1. HISTORY

Suppose A is a C^*-algebra and B is a C^*-subalgebra of A. Theorems of the Stone-Weierstrass type assert that, if some additional conditions are met, then B=A. If M is the set of maximal modular left ideals of A together with A itself, then the original Stone-Weierstrass theorem can be stated as follows. If A is abelian and B separates M (i.e. for $I,J \in M$, $I=J$ if and only if $I \cap B = J \cap B$), then B=A. The general Stone-Weierstrass problem, which remains unsolved, is whether the assumption that A is abelian can be dropped. In this paper we shall review the historical development of the Stone-Weierstrass problem, the partial results, solutions for special cases, solutions with stronger hypotheses and variants. Then we shall discuss how some very recent work of Sorin Popa can be used to reduce the problem further. Full details will be given only in the last section since earlier proofs have already been published.

The original theorem of Weierstrass dealt with uniformly approximating continuous real-valued functions on a closed interval with polynomials on that interval. Stone's generalization replaces the closed interval with an arbitrary compact space Ω and polynomials with any subalgebra of continuous functions on Ω that separates the points of Ω. (See [21] for more details.) The extension to algebras of complex functions on a locally compact Hausdorff space was fairly routine. Because the maximal modular ideals of $C_o(X)$, the algebra of complex continuous functions on the locally compact Hausdorff space X, are all of the form $I_x = \{a \in C_o(X) : a(x) = 0\}$, for $x \in X$, we see that for a subalgebra $B \subset C_o(X)$ to separate the points of X it is necessary and sufficient for B to separate the maximal ideals of $C_o(X)$. This shows that the version of the Stone-Weierstrass theorem stated above is equivalent to Stone's original result.

This identification of the maximal ideals of $C_o(X)$ with the points of X has an analogue in the non-abelian case. If A is a C^*-algebra, there is a one-to-one correspondence between the maximal modular left ideals of A and the pure states of A. This was proved by Kadison in [14]. Thus, if P(A) denotes the set of pure states of A (together with the zero functional if A does not have a unit), the general Stone-Weierstrass problem assumes that B is a C^*-subalgebra of A that separates P(A) and asks if B=A.

Even before Kadison's general result mentioned above, Kaplansky [15] solved the general Stone-Weierstrass problem in the affirmative in the case where A is a C.C.R. algebra. (An algebra is C.C.R. if its image under every irreducible representation consists exactly of the compact operators.) Without great difficulty Kaplansky's ideas extend to give the same result when A is G.C.R., (A is G.C.R. if it has a composition series of ideals such that successive quotients are C.C.R.) and this appears in detail in [9, 11.1.8]. Kaplansky [15] was the first person to conjecture that the result holds in general and, after Kadison's theorem the conjecture seemed even more likely.

In 1959 James Glimm, a recent Ph.D. student of Kadison, began work on a variant of the problem suggested by Kadison. Noting that if A is abelian, the set P(A) is weak*-closed in the dual space A^* of A, Glimm assumed the stronger hypothesis that B separated the weak*-closure $P(A)^-$ of P(A). With this assumption and using some unpublished lemmas of Kadison he showed that B=A [13]. Details can be found also in the book of Dixmier [9, Ch.11]. Although Glimm's proof is both ingenious and deep, his theorem is not entirely satisfactory as a generalization of the Stone--Weierstrass theorem. The reason for this is that the hypothesis no longer has a connection with the left ideal structure of A. In fact Glimm's proof shows that if A is any C^*-algebra reasonably far removed from the G.C.R. case, then $P(A)^-$ consists of the entire state space of A.

Except for the exposition in the book of Dixmier [9] in 1964, which included some previously unpublished simplifications of the Kaplansky and Glimm arguments due to J.M.G. Fell, the Stone--Wierestrass problem received no more attention until the first author of the present paper published some partial results on it in 1969 [1]. This paper attempted several different approaches, some of which were picked up and greatly extended by later authors. It will be helpful to describe the general framework developed in that paper because subsequent results will be easier to present.

We shall consider the C^*-algebras A and B as canonically imbedded in their second duals [16, 3.7.8] A^{**} and B^{**} respectively. If $i:B \to A$ denotes the inclusion map, then $i^{**}:B^{**} \to A^{**}$ can also be regarded as an inclusion map, so we may view B^{**} as a subset of A^{**}. The map $i^*:A^* \to B^*$ is merely the restriction map. To each pure state f on A there corresponds a unique minimal projection p in A^{**} such that $f(p)=1$ and conversely each minimal projection in A^{**} gives rise to a pure state. If z_A denotes the supremum of all minimal projections in A^{**}, then z_A is a central projection in A^{**} and $z_A A^{**}$ is a purely atomic von Neumann algebra, hence isomorphic to a direct sum of algebras of the form B(H). (B(H) denotes all bounded operators acting on the Hilbert space H.) It was shown in [1] that if B separates P(A) and if z_B denotes the analogous projection in B^{**}, then i^{**} is one-to-one and onto from $z_B B^{**}$ to $z_A A^{**}$. Thus, considering B^{**} as a subset of A^{**} we may drop the subscript from z_A and assert that $zB^{**}=$ $=zA^{**}$ and $z \varepsilon B^{**}$. Further, it follows that i^* is isometric from zA^* onto zB^*. To prove the above statement it is necessary to assume that A and B contain the same unit element. If A has no unit, we need only to adjoin a unit to each of A and B. However, it might be, a priori, that A has a unit while B does not. This possibility is eliminated by letting $\{a_\alpha\}$ denote a positive increasing approximate unit for B and noting that $f(1-a_\alpha) \to 0$ for each f in P(A). Since $(1-a_\alpha)$ is a decreasing net in A, it converges in A^{**} to an element that defines a non-negative upper semicountinuous function on the set of states of A which vanishes on P(A). Such a function must be identically zero (see the more general statement in [16, 4.3.15]); so Dini's theorem gives that $(1-a_\alpha) \to 0$ uniformly on the states of A, hence in norm. Thus, $a_\alpha \to 1$ in norm so $1 \in B$.

To describe the first Stone-Weierstrass result from [1] we merely drop the word maximal and get the following theorem. Recall that we can now assume that A and B have the same unit (and from now on we do so).

THEOREM. *If* B *separates the closed left ideals of* A, *then* B=A.

Note that this stronger assumption that B separates all the closed left ideals of A, not just the maximal ideals, does not seem to make the proof of the (abelian) Stone-Weierstrass theorem any easier. Recall that if A is abelian, i.e., $A=C(\Omega)$, the space of continuous, complex-valued functions on a compact Hausdorff space Ω, then the closed ideals of A all have the form $I=\{f \in C(\Omega) : f(\omega)=0 \text{ for all } \omega \in K\}$, where $K \subset \Omega$ is a closed set. The stronger assumption means that if $K \subset \Omega$ is a closed set and ω_o is in $\Omega \setminus K$, then there exists f in B such that $f(\omega_o) \neq 0$ and f is 0 on K.

The above theorem and the analysis of the abelian case suggest that we try to find a non-abelian analogue for the closed subsets of Ω. Such an analogue is supplied in [1]. A projection p in A^{**} is said to be *open* if there exists an increasing net $\{a_\alpha\} \subset A$ of positive elements with $a_\alpha \nearrow p$ in A^{**}. The projection $p' = 1-p$ is said to be *closed*. Note that $(1-a_\alpha) \searrow p'$. As shown more generally in [16, 4.3.15] the map $p \mapsto zp$ is one-to-one on the set of open or closed projections. The closure $\bar{q}$ of a projection q in A^{**} is by definition the smallest closed projection in A^{**} that majorizes q. In the abelian case a projection q in zA^{**} exactly corresponds to a set in Ω (where $A=C(\Omega)$, Ω compact Hausdorff) and its closure $\bar{q}$ corresponds to the closure of the set in Ω. Since it follows easily from the 1963 work of Effros [10] that the formula $I=\{a \in A : aq=0\}$ defines a bijection between closed left ideals in A and closed projections in A^{**}, to show that B separates the closed left ideals of A (and hence B=A) we need to show that every closed projection in A^{**} actually lies in B^{**}.

To do this let p be a closed projection in A^{**}, set $p_o=zp$ and recall that $zA^{**}=zB^{**}$ so $p_o \varepsilon B^{**}$. By [1, II.16] p is the closure in A^{**} of p_o. Since p_o lies in B^{**} it makes sense to let q denote the closure of p_o *in* B^{**} and to try to show that q=p. (Clearly q is closed in A^{**} so that $q \geq p$.) Thus, the heart of the matter is to show that A and B generate the same "topology".

Using the above notation the problem has been reduced to showing that p=q, but a further reduction is possible using the notion of regularity of projections introduced by Tomita [26] and expanded by Effros [11]. As in [1, II.11] we define a projection r in A^{**} to be *regular* for A if $||ar||=||a\bar{r}||$ for every a in A. Since p_o (notation as above) lies in zA^{**}, it makes sense to consider it in B^{**} also, so we might ask if it is regular for B. If p_o is regular for B, then it can be shown that p=q using [10, pp.408-9]. Regularity was equivalently defined by Tomita and Effros as follows. A projection r in A^{**} is regular if

$$||ar||=\inf\{||a+b||: b \varepsilon A,\ br=0\}.$$

In fact to show that p_o is regular for B we need only take a net $\{a_\alpha\}$ in A with $1 \geq a_\alpha \searrow p$ in A^{**} and try to show that for each b in B,

$$\lim_\alpha ||ba_\alpha||=\inf\{||b+c|| \ : \ c \varepsilon B,\ \lim_\alpha ||ca||=0\}.$$

Unfortunately, we can go no further along this line at this time.

Another approach in [1], which was subsequently improved by other authors, was to postulate the existence of an inverse to one of the maps i, i^* or i^{**}. For example in [1, III.9] it is shown that if there exists a norm one projection φ from A onto B, then B=A. Of course $\varphi \circ i$ is the indentity on B, so $i^* \circ \varphi^*$ is the identity on B^*. It is straightforward to check that φ^* must be one-to-one and onto from the unit ball of zB^* to the unit ball of zA^*, and a continuity argument then shows that φ^* maps the unit ball of B^* onto the unit ball of A^*. Since φ^* is onto, i^* must be one-to-one, so i is onto and hence B=A.

Later Effros [11] proved a more general result which applied to operator systems (one example of which is the real subspace of self-adjoint elements in a C^*-algebra). When his result is specialized to the situation of the present paper, Effros assumes the existence of a positive right inverse D for i^* but does not require that D be weak*-continuous as in the previous paragraph. As noted by the second author and John Bunce in [3, Corollary 4] $D^*: A^{**} \to B^{**}$ is a norm one projection and, together with [1, III.7] and a result of Sakai [18] (to be described below), that is enough to conclude that B=A when A is separable.

As an application of the techniques described above, we know that if B=B(H), the algebra of all bounded operators on a Hilbert space H, then there is a norm one projection from A onto B [4, Theorem 1.2.3], thus A=B as above. (See also [20]). Also if A is a semi-finite von Neumann algebra and B is a weakly closed subalgebra of A, then B=A (see [2,II.8] and [1, III.10.(4)].)

In 1970 Sakai gave a substantial extension of Kaplansky's theorem by showing that B=A whenever A is separable and B is nuclear [18, p.393]. Today it is possible to give a quick proof of this result using Effros' theorem and the fact that B is nuclear if and only if B^{**} is injective. At the time, however, this deep fact was not known and Sakai's proof relied on the following theorem.

THEOREM. *Suppose* A *is separable,* B *separates* P(A), *and* $\pi: A \to B(H_\pi)$ *is a representation of* A *on separable Hilbert space. If* $\mathcal{A}$ *is a maximal abelian subalgebra of* $\pi(A)'$ *and* φ *is a linear norm one map of* $\pi(A)$ *into* $W^*(\pi(A), \mathcal{A})$ *(the von Neumann algebra generated by* $\pi(A)$ *and* $\mathcal{A}$*) which is the identity on* $\pi(B)$, *then* φ *is the identity on* $\pi(A)$.

Thus φ could be a projection of norm one from A to B, but in general its range is allowed to be much larger. This theorem is used extensively by the second author and John Bunce in [3].

In a 1980 paper [5] John Bunce assumes that the map i has a weak approximate left inverse and (assuming A to be separable) concludes that B=A. Specifically, he assumes that there exists a sequence (not a net) $\{L_n\}$ of norm one linear maps from A to B such that $(L_n \circ i)(b) \to b$ in the $\sigma(B^*, B)$ topology for all b in B.

The only other published theorem of the Stone-Weierstrass type up through 1980 is a short note by Elliot in which he proves B=A under the additional assumption that A is generated by B and a single element a in A satisfying $ab-ba\in B$ for all b in B. No separability is required and the argument resembles Sakai's proof [19] that every derivation of a simple C^*-algebra with unit is inner.

2. THE FACTORIAL STONE-WEIERSTRASS PROBLEM AND THE WORK OF POPA

Throughout this section we assume that A is a separable C^*-algebra, B is a C^*-subalgebra and B contains the unit of A. Recall that a state f on A is said to be *factorial* (or a *factor state*) on A if $\pi_f(A)''$ is a factor, where π_f denotes the cyclic representation of A induced by f, [16, p.46]. If A is abelian, then f is factorial if and only if f is pure. Thus, it is possible that the correct non-commutative Stone-Weierstrass theorem should require that B separate F(A), the set of factor states of A. Some results concerning this factorial Stone-Weierstrass conjecture were obtained by the second author and John Bunce in [3]. Moreover, using some recent work of Sorin Popa [17] it is possible to reduce the (separable) factorial Stone-Weierstrass problem to a question concerning type III_1 factors. Our purpose in this section is to present this reduction. We begin by recapitulating some of the results of [3].

THEOREM 1. ([3, Theorem 12]). *If* B *separates* F(A) *and each factor state of* B *extends to a factor state of* A, *then* B=A.

Recall that if A is a maximal abelian sublagebra of a factor M, a unitary u in M *normalizes* A if $uAu^*=A$. By definition A is *semiregular* if the unitaries in M that normalize A generate a factor.

THEOREM 2. ([3, Proposition 13]). *If* B *separates* F(A), $f\in F(A)$ *and* $\pi_f(B)'$ *contains a semiregular maximal abelian subalgebra, then* f *extends to a factor state on* A.

We remark that in fact it is enough that $\pi_f(B)''$ contain a semiregular maximal abelian subalgebra in order that f extend to a factor state on A. (See the proof of [3, Theorem 10].)

Thus, if each separable factor contains a semiregular maximal abelian subalgebra, then the factorial Stone-Weierstrass conjecture is true. The following result is due to Sorin Popa [17, Theorem 3.2]. We say that a von Neumann algebra is *separable* if it has a separable faithful representation.

THEOREM 3. *If M is a separable factor and N is a semifinite subfactor of M satisfying*

i) $N' \cap M = \mathbb{C}$ *and*

ii) *there is a normal conditional expectation of M onto N, then N contains a semiregular maximal abelian subalgebra of M.*

Hence if $f \in F(B)$ and $\pi_f(B)''$ is semifinite, then $\pi_f(B)'$ is semifinite and (taking $N=M=\pi_f(B)'$) it follows from Theorems 2 and 3 that f extends to a factor state on A. By using the work on the classification of factors done by Connes and Takesaki we may derive a similar result for type III_λ representations $(0 \leq \lambda < 1)$.

The main tool to be used below is the discrete decomposition for factors of type III_λ $(0 \leq \lambda < 1)$ which was developed by Connes in his thesis [6]. We now describe the salient features of this decomposition. Suppose M is a factor of type III_λ with $0 \leq \lambda < 1$. Connes showed that M is the crossed product of a type II_∞ algebra by a single automorphism. That is, M is generated by a II_∞ subalgebra N and a single unitary element. If $\lambda > 0$ we may take N to be a factor with $N' \cap M = \mathbb{C}$. Moreover, there is a normal conditional expectation of M onto N. If $\lambda = 0$, then N has diffuse center. (An abelian von Neumann algebra is diffuse if it has no minimal projections.) Moreover, if u denotes the unitary element mentioned above then the automorphism ad_u maps the center into itself and acts ergodically on it. Finally each maximal abelian subalgebra of N is maximal abelian in M . These results are contained in Sections IV and V of Connes' thesis. For another treatment see [22; Sections 29, 30].

If $f \in F(B)$ and $\pi_f(B)''$ is of type III_λ with $0<\lambda<1$, then $\pi_f(B)'$ is also of type III_λ and so Theorems 2 and 3 and the discrete decomposition give that f extends to a factor state in this case also. In sum Popa's theorem and the work of Connes yield the following result.

THEOREM 4. *If* B *separates* F(A), $f \in F(B)$ *and* $\pi_f(B)''$ *is a factor that is not of type* III_0 *or* III_1 *, then* f *extends to a factor state on* A.

The remainder of this section is devoted to dealing with the type III_0 case. This will be accomplished by proving a slight generalization of Popa's theorem and using the properties of the discrete decomposition. The following generalization of Popa's theorem will be sufficient for our needs.

THEOREM 5. *If* N *is a separable type* II_1 *von Neumann algebra with center* Z*, then there is a maximal abelian subalgebra* A *and a hyperfinite von Neumann algebra* R *such that* $A \subset R \subset N$*,* R *is generated by unitaries that normalize* A *and* $R' \cap N = Z$.

We postpone the proof for the moment and next show how Theorem 5 can be used to handle the type III_0 case.

THEOREM 6. *If* B *separates* F(A), $f \in F(B)$ *and* $\pi_f(B)''$ *is a type* III_0 *factor, then* f *extends to a factor state on* A.

PROOF. Write $M=\pi_f(B)'$ so that M is a type III_0 factor. Apply the discrete decomposition to M to obtain a type II_∞ algebra P with diffuse center Z and a unitary element u such that P and u generate M and ad_u acts ergodically on Z. By [24, Chap.V, 1.40] we may write $P=N \otimes B(H)$, where H is separable and N is a II_1 von Neumann algebra with center Z_1. Note that $Z=Z_1 \otimes 1$. Fix an orthonormal basis $\{\eta_n : n \in \mathbb{Z}\}$ for H and let S denote the bilateral shift along this basis ($S\eta_n = \eta_{n+1}$). Also let $\mathcal{D}$ denote the algebra of diagonal operators with respect to this basis. We have then that $\mathcal{D}$ is maximal abelian in B(H) and S normalizes $\mathcal{D}$. Apply Theorem 5 to N to get A and R as above. If we write $\mathcal{B}=A \otimes \mathcal{D}$ and $\mathcal{S}=\{R \otimes 1, 1 \otimes S, 1 \otimes \mathcal{D}\}''$ then $\mathcal{B}$ is maximal abelian in P, $\mathcal{S}$ is generated by unitaries that normalize $\mathcal{B}$ and $\mathcal{S}'\cap\mathcal{P}=\mathcal{Z}$. Indeed since S and $\mathcal{D}$ generate B(H), $\mathcal{S}'\cap\mathcal{P}=\mathcal{R}'\otimes 1\cap\mathcal{N}\otimes B(H)=\mathcal{Z}_1\otimes 1=\mathcal{Z}$. As noted above $\mathcal{B}$ is maximal abelian in $\mathcal{M}$ and so $\mathcal{S}'\cap\mathcal{M}=\mathcal{Z}$. By [3, Theorem 6] there is a representation π of A such that $\pi(A)$ acts on the same Hilbert space as $\pi_f(B)$ and such that

i) π extends π_f ,

ii) $\mathcal{B}\subset\pi(A)'$,

iii) $\pi(A)''\subset W^*(\pi_f(B),\mathcal{B})$ and

iv) π is the unique representation that satisfies i) and ii).

To prove the theorem it suffices to show $\pi(A)''$ is a factor; for then the cannonical cyclic vector will give the desired factor state extension of f. If v is a unitary in $\mathcal{M}$ that normalizes $\mathcal{B}$, then the map ρ defined by $\rho(a)=v\pi(a)v^*$ is a representation of A that satisfies i) and ii) and so $\rho=\pi$, and $v\in\pi(A)'$. Hence $\mathcal{S}\subset\pi(A)'$ and therefore $\pi(A)''\cap\pi(A)'\subset\mathcal{S}'\cap\mathcal{M}=\mathcal{Z}$. As B separates F(A) we have by [3, Proposition 14] that $\pi(A)''\subset W^*(\pi_f(B),\mathcal{Z})$. As ad_u normalizes $\mathcal{Z}$ it follows as above that $u\in\pi(A)'$. On the other hand ad_u acts ergodically on $\mathcal{Z}$. Hence the center of $\pi(A)''$ is trivial and the theorem is proved.

It only remains to prove Theorem 5. It is convenient to break the proof up into a series of lemmas. We begin by introducing some notation and recalling some facts. Fix a type II_1 von Neumann algebra N and let τ denote a faithful normal trace on N. We identify N with its image under the representation induced by τ so that N acts on a Hilbert space H_τ and H_τ contains a cyclic vector η_τ that gives the trace. (For x in N, $\tau(x)=(x\eta_\tau,\eta_\tau)$ and $\tau(x^*x)=||x||_\tau^2$.) If A is any von Neumann subalgebra of N there is a faithful normal conditional expectation E_A of N onto A that is given as follows. If P_A denotes the projection of H_τ onto $[A\eta_\tau]$ and Φ the map of N into $B(P_AH_\tau)$ defined by

$$\Phi(x)=P_Ax|P_AH_\tau ,$$

then $\Phi(x)\in AP_A$, Φ is injective on A and E_A is defined by

$$E(x)=\Phi^{-1}(P_Ax|P_AH_\tau).$$

The map E_A preserves τ; $\tau=\tau\circ E_A$ and is the unique normal conditional expectation with this property. (See [22, Theorem 10.1].) Our proof relies on two key lemmas of Popa which we record below.

LEMMA 7. ([17, 1.2]). *If* $\{A_i\}$ *is an increasing sequence of abelian subalgebras of* N *and* $A=\{\bigcup_i A_i\}''$, *then* A *is maximal abelian in* N *if and only if*

$$\lim_i ||E_{A_i'\cap N}(x)-E_{A_i}(x)||_\tau=0$$

for each x *in* N.

LEMMA 8. ([17, 2.5]). *If* $x_1,\ldots,x_n \varepsilon N$ *and* $\varepsilon>0$, *then there is a finite-dimensional abelian subalgebra* A *of* N *such that*

$$||E_{A'\cap N}(x_i)-E_A(x_i)||_\tau<\varepsilon||x_i||_\tau\ , \qquad i=1,\ldots,n.$$

If A is a finite-dimensional abelian subalgebra of N, then it is easy to check that for x in N

$$E_A(x)=\sum_i(\tau(e_i x e_i)/\tau(e_i))e_i$$

and

$$E_{A'\cap N}(x)=\sum_i e_i x e_i\ ,$$

where the e_i's denote the minimal projections of A. Using these formulas the following fact is also easy to show and we omit its proof.

PROPOSITION 9. *If* A *is a finite-dimensional abelian subalgebra of* N *with minimal projections* $e_1,\ldots,e_n$ *and* $\varepsilon>0$, *then there is* $\delta>0$ *such that if* $f_1,\ldots,f_n$ *are projections in* N *with* $f_i\leq e_i$ *and* $\sum_i||e_i-f_i||_\tau<\delta$, *then for each* x *in* N

$$||E_{A'\cap N}(x)-E_{B'\cap N}(x)||_\tau+||E_A(x)-E_B(x)||_\tau\leq\varepsilon||x||_\tau\ ,$$

where B *denotes the algebra generated by* 1 *and* $f_1,\ldots,f_n$.

The proof of Theorem 5 will use the notion of a fundamental

projection. Recall [8, III, 8.2] that a projection e in a von Neumann algebra N is *fundamental* if there is a central projection z and equivalent mutually orthogonal projections $e_1,\dots,e_{2^n}$ with sum z and such that $e=e_1$. If N is of type II_1, then each projection is the (possibly infinite) sum of mutually orthogonal fundamental projections [8, Corollaire, III, 8.2]. Let us say that a finite-dimensional abelian subalgebra of A is *fundamental* if it is generated by a set of minimal projections of the form $\{e_j^{(i)}: 1\le j\le 2^n,\ 1\le i\le m\}$ where for each fixed i, the $e_j^{(i)}$ are equivalent and sum to a central projection z_i. Note that if A and B are abelian subalgebras of N and $B\subset A$ then for each x in N

$$||E_{A'\cap N}(x)-E_A(x)||_\tau^2=||(P_{A'\cap N}-P_A)x\eta_\tau||_\tau^2\le$$
$$\le||(P_{B'\cap N}-P_B)x\eta_\tau||_\tau^2=||E_{B'\cap N}(x)-E_B(x)||_\tau^2.$$

PROPOSITION 10. *If* $x_1,\dots,x_n$ *are in* N *and* $\varepsilon>0$ *then there is a fundamental finite-dimensional abelian subalgebra* A *of* N *such that*

$$||E_{A'\cap N}(x_i)-E_A(x_i)||_\tau\le\varepsilon||x_i||_\tau\ ,\qquad i=1,\dots,n\ .$$

PROOF. By Lemma 8 there is a finite-dimensional abelian subalgebra B of N such that

$$||E_{B'\cap N}(x_i)-E_B(x_i)||_\tau\le(\varepsilon/2)||x_i||_\tau\qquad i=1,\dots,n\ .$$

Each of the minimal projections in B is the (possibly infinite) sum of mutually orthogonal fundamental projections. By replacing each minimal projection by a smaller one if necessary and invoking Proposition 9 we obtain an algebra C such that each of its minimal projections is the sum of a *finite* number of fundamental mutually orthogonal projections and

$$||E_{C'\cap N}(x_i)-E_C(x_i)||_\tau\le\varepsilon||x_i||\qquad i=1,\dots,n\ .$$

The central projections associated to these fundamental projections may not be orthogonal; however, by taking appropriate products of these projections we get orthogonal central projections $z_1,\ldots,z_m$ which sum to 1 and such that the algebra they generate contains the original ones. By considering products of the z_i's and the fundamental projections and halving when necessary, we get a family $\{e_j^{(i)}:1\le j\le 2^k,\ 1\le i\le m\}$ such that for each fixed i the $e_j^{(i)}$'s sum to z_i and are equivalent and such that if A denotes the algebra they generate, then $B\subset A$. By the remark preceding the statement of the proposition A has the desired properties.

PROOF OF THEOREM 5. Fix a sequence $\{x_n\}$ in N that is dense in N in trace norm. We shall construct by induction sequences A_k and R_k such that each sequence is increasing, each A_k is a fundamental finite-dimensional abelian subalgebra of N and each R_k is a finite-dimensional matrix subalgebra of N whose diagonal matrix units are the minimal projections of A_k. Moreover each R_k will be generated by unitaries that normalize all of the A_j's and the fixed points under the action of these unitaries will be precisely the central projections of N contained in the A_j's. Suppose that the first k of each of these algebras have been constructed such that

$$||E_{A_k'\cap N}(x_n)-E_{A_k}(x_n)||_\tau\le 2^{-k}||x_n||_\tau \qquad 1\le n\le k\ .$$

To avoid notational complexity we make some reductions. Since each direct summand of N given by a minimal central projection of R_k can be considered separately, we assume that the center of R_k is trivial so that A_k is generated by the equivalent minimal projections $e_1,\ldots,e_r$, where r is a power of 2. Also suppose that R_k is generated by partial isometries $v_1,\ldots,v_r$ such that

$$v_1=e_1\ ,\ v_jv_j^*=e_j\ ,\ v_j^*v_j=e_1 \qquad \text{for } j=2,\ldots,r\ .$$

Write $e=e_1$, $\tau_e=\tau(e\cdot)/\tau(e)$ and $N_e=eNe$. Apply Proposition 10 to the algebra N_e and the elements $v_j^*x_nv_j$ for $j=1,\ldots,r$ and $n=1,\ldots,k+1$ and $\varepsilon=2^{-(k+1)}$ to obtain projections $f_q^{(p)}$, $q=1,\ldots,s$;

$p=1,\ldots,m$; s is another power of 2, such that for each fixed p the $f_q^{(p)}$'s are equivalent and sum to a central projection in N_e. Thus if we let A_e denote the algebra generated by the $f_q^{(p)}$'s, then we have

$$||E_{A_e' N_e}(v_j^* x_n v_j)-E_{A_e}(v_j^* x_n v_j)||_{\tau_e} \leq 2^{-(k+1)}(||e_j x_n e_j||_\tau/||e||_\tau).$$

Select partial isometries $v_q^{(p)}$ in N_e such that $v_1^{(p)}=f_1^{(p)}$, and for $q=2,\ldots,s$; $p=1,\ldots,m$, $v_j^{(p)}(v_j^{(p)})^*=f_j^{(p)}$, $(v_j^{(p)})^* v_1^{(p)}=f_1^{(p)}$. Finally let A_{k+1} denote the algebra generated by the projections of the form

$$v_j f_q^{(p)} v_j^*$$

and by R_{k+1} the algebra generated by the partial isometries of the form

$$v_j v_q^{(p)} v_j^* \quad .$$

It is straightforward to check that

$$||E_{A_{k+1}'\cap N}(x_n)-E_{A_{k+1}}(x_n)||_\tau^2=$$

$$=\sum_j ||E_{A_{k+1}'\cap N}(v_j^* x_n v_j)-E_{A_{k+1}}(v_j^* x_n v_j)||_\tau^2=$$

$$=\sum_j ||E_{A_e'\cap N_e}(v_j^* x_n v_j)-E_{A_e}(v_j^* x_n v_j)||_{\tau_e}^2 \, ||e||_\tau^2 \leq$$

$$\leq (2^{-(k+1)})^2 ||x_n||_\tau^2 \quad .$$

Also it is easy to see that if w is a central projection in N_e then the formula

$$z=\Sigma v_j w v_j^*$$

determines a central projection in N. It follows that A_{k+1} is a fundamental finite-dimensional subalgebra of N and R_{k+1} is a finite-dimensional matrix algebra with the required properties.

It also follows that for each x in N

$$||E_{A'_k \cap N}(x) - E_{A_k}(x)||_\tau \to 0 \qquad \text{as } k \to \infty ,$$

and so by Lemma 7 the weak closure A of the A_k's is maximal abelian in N. If we write R for the algebra generated by the R_k's, then it is clear that R is hyperfinite and generated by unitaries that normalize R. To establish the final assertion of the Theorem let E_k denote the conditional expectation of N onto R_k and fix x in $R' \cap N$. As A is maximal abelian we have that x is in A and therefore in R. Also it is clear from the definition of the conditional expectation given above that $E_k(x)$ converges to $E(x)$ in trace norm, where E denotes the conditional expectation of N onto R. Since $E_k(x)y=E_k(xy)=yE_k(x)$ for all y in R_k, $E_k(x)$ lies in the center of R_k and hence in the center of N. Thus, $E(x)$ lies in the center of N. Since x is in R, $E(x)=x$ and so x is in the center of N.

In the III_1 case we no longer have the discrete decomposition. However, it is known that if M is a type III_1 factor, then M has a continuous decomposition. That is M is the crossed product of a II_∞ von Neumann algebra by a one-parameter group of automorphisms. Moreover, this subalgebra may be chosen so that it is a factor with trivial relative commutant ([7, 25]). By Popa's theorem, the subalgebra contains a semiregular maximal abelian subalgebra A. However, it is not known it A is maximal abelian in M. In order to apply Popa's methods to the III_1 case it seems necessary to find a faithful normal state on M such that its centralizer has trivial relative commutant. Connes has shown (in unpublished work) that the existence of such a state would imply the uniqueness of the separable hyperfinite III_1 factor, a problem that thus far has defied determined attack by experts.

We conclude by mentioning some directions for future research. In the abelian case the relationship between a C^*-algebra A and its C^*-subalgebra B is completely determined by the fact that the maximal ideal space of B must be a quotient of the maximal ideal space of A. Thus any algebraic question immediately becomes a topological one. The richer set of possibilities in the non-abelian case does not allow us to make direct use of such a correspondence, except occasionally by analogy. In particular, the existence of factor states that are not pure leads to many fascinating questions about C^*-algebras. For example, does every factor state of a C^*-subalgebra of A extend to a factor state of A? As we see from Theorem 1 above, an affirmative solution would prove the factorial Stone-Weierstrass conjecture, and it would have other important consequences as well. However, except for the type I case (in which all factor representations are type I), this extension problem is open. A related question would be to describe the class of (separable) C^*-algebras that have no type III_1 factor representations. This may be exactly the class of type I C^*-algebras, but, again no work on this question has appeared.

For those researchers who would like to try to make some progress on the Stone-Weierstrass conjecture, here are a few preliminary suggestions.

1. Assume that the larger algebra is separable and nuclear. It is known that subalgebras of nuclear C^*-algebras need not be nuclear, so Sakai's theorem does not apply immediately. However, the additional assumption that the subalgebra separates the pure (or factor) states may be enough to prove that it is nuclear.

2. Concentrate on a particular algebra A, for example a group C^*-algebra or an infinite tensor product, where the pure states are somewhat better understood. In particular let A be the Fermion algebra [16, p.206] or the C^*-algebra arising from the left regular representation of the free group on two generators. See [7, 13.6 and 13.9] for details on group C^*-algebras and their pure states.

3. Study the compact subsets of pure states of a C^*-algebra. This is of interest since by Choquet theory each state on a C^*-algebra A has a regular representing measure that is concentrated on the pure states of A.

REFERENCES

1. Akemann, C.A.: The general Stone-Weierstrass problem, *J.Functional Analysis* 4(1969), 277-294.

2. Akemann, C.A.: Left ideal structure of C*-algebras, *J.Functional Analysis* 6(1970), 305-317.

3. Anderson, J.; Bunce, J.: Stone-Weierstrass theorems for separable C*-algebras, *J.Operator Theory* 6(1981), 363-374.

4. Arveson, W.B.: Subalgebras of C*-algebras, *Acta Math.* 123 (1969), 141-244.

5. Bunce, J.W.: Approximating maps and a Stone-Weierstrass theorem for C*-algebras, *Proc.Amer.Math.Soc.* 79(1980), 559--563.

6. Connes, A.: Une classification des facteurs de type III, *Ann.Ec.Norm.Sup.* 6(1973), 133-252.

7. Connes, A.; Takesaki, M.: The flow of weights on a factor of type III, *Tôhoku Math.J.* 29(1977), 473-575.

8. Dixmier, J.: *Les algèbres d'opérateurs dans l'espace hilbertien (Algèbres de von Neumann)*, 2d ed., Gauthier-Villars, Paris, 1969.

9. Dixmier, J.: *Les C*-algèbres et leur représentations*, Gauthier-Villars, Paris, 1964.

10. Effros, E.G.: Order ideals in a C*-algebra and its dual, *Duke Math.J.*, 30(1963), 391-412.

11. Effros, E.G.: *Injectives and tensor products for convex sets and C*-algebras*, Lectures given at the NATO Institute on "Facial structure of compact convex sets" at the University College of Swansea, Wales, July, 1972.

12. Elliot, G.A.: Another weak Stone-Weierstrass theorem for C*-algebras, *Canad.Math.Bull.* 15(1972), 355-358.

13. Glimm, J.: A Stone-Weierstrass theorem for C*-algebras, *Ann. of Math.* 72(1960), 216-244.

14. Kadison, R.V.: Irreducible operator algebras, *Proc.Nat.Acad. Sci.* 43(3)(1957), 273-276.

15. Kaplansky, I.: The structure of certain operator algebras, *Trans.Amer.Math.Soc.* 70(1951), 219-255.

16. Pedersen, G.K.: *C*-algebras and their automorphism groups*, Academic Press, London, 1979.

17. Popa, S: On a problem of R.V.Kadison on maximal abelian *-subalgebras in factors, *Invent.Math.* 65(1981), 269-281.

18. Sakai, S.: On the Stone-Weierstrass theorem for C*-algebras, *Tôhoku Math.J.* 22(1970), 191-199.

19. Sakai, S.: Derivations of simple C*-algebras, *J.Functional*

Analysis 2(1968), 202-206.

20. Schultz, F.: Pure states on a dual object for C*-algebras, *Comm.Math.Phys.*, to appear.

21. Stone, M.H.: The generalized Weierstrass approximation theorem, *Math.Mag.* 21(1948), 167-184, 237-254.

22. Strătilă, S.: *Modular theory in operator algebras*, Abacus Press, Tunbridge Wells, 1981.

23. Takesaki, M.: Conditional expectation in von Neumann algebra, *J.Functional Analysis* 9(1972), 306-321.

24. Takesaki, M.: *Theory of operator algebras. I*, Springer-Verlag, New York, Heidelberg, Berlin, 1979.

25. Takesaki, M.: Duality for crossed products and the structure of von Neumann algebras of type III, *Acta Math.* 131(1973), 249-310.

26. Tomita, M.: Spectral theory of operator algebras. I, *Math.J. Okayama Univ.* 9(1959), 63-98.

C.Akemann
Department of Mathematics,
University of California,
Santa Barbara, CA 93106,
U.S.A.

J.Anderson
Department of Mathematics,
Pennsylvania State University,
University Park, PA 16802,
U.S.A.

Both authors were supported in part by a grant from the National Science Foundation.

ON THE NORM-CLOSURE OF THE SIMILARITY ORBIT OF ESSENTIALLY NILPOTENT OPERATORS

Constantin Apostol

INTRODUCTION

Let T be a bounded linear operator acting in a complex, separable Hilbert space and let $S(T)$ denote its similarity orbit, i.e. $S(T)=\{A^{-1}TA\}$. D.A. Herrero, [5], raised the problem of characterizing $S(T)^{-}$, the norm-closure of $S(T)$, in terms of the spectral properties of T. The problem is still open for essentially nilpotent operators. If T is an essentially nilpotent operator, we do not know which are the natural necessary conditions, for S to belong to $S(T)^{-}$, strong enough to be sufficient, too.

The aim of this paper is to put in evidence two necessary conditons for S to belong to $S(T)^{-}$, if T is an essentially nilpotent operator. They will be called below, the *index condition* (Theorem 2.3) and resp. the *range condition* (Theorem 3.3). The index condition and Proposition 1.4 and Proposition 1.5 represent a new presentation of some facts known by D.Voiculescu and the author since 1977, to be published in the monograph [6], Section 8.4.3 (see [6], 8.6, Notes and remarks). The matrix representation of essentially nilpotent operators, described in Section 1, Theorem 1.2, is the new tool we use to derive the index condition in Section 2 and the range condition in Section 3. The range condition represents the original part of the paper and its unexpected content could be taken as a mark of the difficulty of the problem. The feeling of the author is that there exists a third natural condition only, involving "the rank", whose statement represents the last mystery of the problem.

We mention that, in fact, the essentially nilpotent operators T for which $S(T)^{-}$ is not known are some particular individuals in the class, which will be precisely described in Section 1.

1. NOTATION AND PRELIMINARIES

Throughout the paper we shall denote by H, a complex, separable, infinite-dimensional Hilbert space, by $L(H)$, the algebra of all bounded linear operators acting in H and by $K(H)$, the ideal in $L(H)$ of all compact operators. For every $T\varepsilon L(H)$, the symbol $\tilde{T}$ will denote the image of T in the Calkin algebra $L(H)/K(H)$. The operators $T\varepsilon L(H)$, $T'\varepsilon L(H')$ will be called *similar* if there exists an invertible operator $A\varepsilon L(H',H)$ such that $T'=A^{-1}TA$ and the similarity orbit of T will be the set

$$S(T)=\{S\varepsilon L(H):\ S \text{ is similar with } T\}.$$

Further we introduce some notation borrowed from [6].

q_m: the canonical truncated shift acting in $\mathbb{C}^m(q_1=0)$, $m\geq 1$;

$q_m^{(k)}$: the orthogonal sum of k copies of q_m, $1\leq k\leq\infty$;

$N_m^+(H)=\{T\varepsilon L(H):\ \tilde{T}^m=0,\ \tilde{T}^{m-1}+\tilde{T}^* \text{ is invertible}\}$, $m\geq 1$;

$N_{m,k}^+(H)=K(H)+\{T\varepsilon L(H):\ T \text{ is similar with } q_1^{(k)}\oplus q_m^{(\infty)}\}$,
$m\geq 1$, $0\leq k\leq m-1$;

$S^+(T)=S(T)^-\cap N_m^+(H)$, $T\varepsilon N_m^+(H)$, $m\geq 1$.

The difficult case to be handled, which we mentioned in the introduction is when $T\varepsilon N_m^+(H)$ or more precisely the description of $S^+(T)$ for $T\varepsilon N_m^+(H)$.

In this section we shall give some basic properties of the operators $T\varepsilon N_m^+(H)$.

1.1. LEMMA. *Let* $m\geq 1$, $T\varepsilon N_m^+(H)$ *be given and let* X *be a subspace in* H *such that* $T^{m-1}|X$ *is essentially bounded from below and* $T^{m-1}|X^\perp$ *is compact. Define*

$$T_X^{(m)}:\underbrace{X\oplus\ldots\oplus X}_{m\text{-}times}\to H$$

by the equation

$$T_X^{(m)}\Big(\bigoplus_{k=1}^{m}x_k\Big)=\sum_{k=1}^{m}T^{m-k}x_k,\qquad x_k\varepsilon X.$$

Then $T_X^{(m)}$ *is Fredholm.*

PROOF. Let $\{\bigoplus_{k=1}^{m} x_{k,n}\}_{n=1}^{\infty}$ be a weakly convergent to 0 sequence such that $\lim_{n\to\infty}||T_X^{(m)}(\bigoplus_{k=1}^{m} x_{k,n})||=0$. Since $T^m\varepsilon K(H)$ and $T^{m-1}|X$ is essentially bounded from below and we have

$$\lim_{n\to\infty}||T^{m-1}x_{m,n}||=\lim_{n\to\infty}||T^{m-1}T_X^{(m)}(\bigoplus_{k=1}^{m} x_{k,n})||=0$$

we derive $\lim_{n\to\infty}||x_{m,n}||=0$. But proceeding analogously we deduce in m steps, $\lim_{n\to\infty}||\bigoplus_{k=1}^{m} x_{k,n}||=0$ and this implies that $T_X^{(m)}$ is essentially bounded from below. Let P denote the orthogonal projection of H onto $\ker(T_X^{(m)})^*$. Using the relations

$$PH=(\operatorname{ran} T_X^{(m)})^{\perp}\subset X^{\perp},\ T^m\varepsilon K(H),$$

$$\langle T^m x_1,Ph\rangle=\langle\sum_{k=1}^{m} T^{m-k}x_k\ ,T^*Ph\rangle,\ x_k\varepsilon X,\ h\varepsilon H$$

we derive succesively $(I-P)T^*P\varepsilon K(H)$, $(PT^*P)^m\varepsilon K(H)$, $T^{m-1}P\varepsilon K(H)$. If $\dim PH=\infty$ it is easy to see that $T^*|PH$ can not be essentially bounded from below, contradicting the fact that $T^{m-1}+T^*$ is Fredholm and $T^{m-1}|PH$ is compact. It follows that PH has finite dimension and the proof is concluded.

For every $m\geq 1$, $T\varepsilon N_m^+(H)$, $\varepsilon>0$ put

$$H^{(m)}=H\oplus\ldots\oplus H,\qquad m \text{ times}$$

$$N_m^+(T,\varepsilon)=\{S\varepsilon N_m^+(H):||S-T||<\varepsilon\}\ .$$

1.2. THEOREM. *Let* $m\geq 2$, $T\varepsilon N_m^+(H)$ *be given. Then there exist* $\varepsilon>0$ *and a continuous function*

$$F:N_m^+(T,\varepsilon)\to L(H^{(m)},H)$$

such that $F(S)$ *is invertible,* $S\varepsilon N_m^+(T,\varepsilon)$ *and*

$$F(S)^{-1}SF(S)=\begin{bmatrix} S_{1,1} & S_{1,2} & 0 & \cdots & & 0 \\ S_{2,1} & 0 & I & 0 & \cdots & 0 \\ \vdots & & & & & \\ & & & & & I \\ S_{m,1} & 0 & \cdots & & & 0 \end{bmatrix},$$

where $S_{1,2}$ *is a Fredholm isometry and* $S_{k,1}\varepsilon K(H)$, $1\le k\le m$.

PROOF. Using the relations (for $S\varepsilon N_m^+(H)$)

$$(\tilde{S}^{m-1}+\tilde{S}^*)^*(\tilde{S}^{m-1}+\tilde{S}^*)=\tilde{S}^{*m-1}\tilde{S}^{m-1}+\tilde{S}\tilde{S}^*\ ,\qquad (\tilde{S}^{*m-1}\tilde{S}^{m-1})(\tilde{S}\tilde{S}^*)=0$$

we derive easily that 0 is an isolated point in $\sigma(\tilde{S}^{*m-1}\tilde{S}^{m-1})$. Let $a>0$, $\eta>0$ be such that, for every $S\varepsilon N_m^+(T,\eta)$ we have

$$a\varepsilon\rho(S^{*m-1}S^{m-1}),\quad [0,a]\cap\sigma(\tilde{S}^{*m-1}\tilde{S}^{m-1})=\{0\},\quad \|E(S)-E(T)\|<1$$

where $E(S)$ denotes the spectral projection of $S^{*m-1}S^{m-1}$ corresponding to $[a,\infty)$. If we put $X(S)=E(S)H$, then we know that the operator $S_{X(S)}^{(m)}$, defined in Lemma 1.1, is Fredholm. Let $U,V\varepsilon L(H)$ be isometries such that $UH\subset X(T)$, $\dim(X(T)\ominus UH)<\infty$, $T_{UH}^{(m)}$ is injective and $\ker V^*=\ker(T_{UH}^{(m)})^*$. Now if we denote by $U(S)$, $V(S)$ the partial isometries determined by the polar decomposition of $E(S)U$, resp. $P_{\mathrm{ran}S_{U(S)H}^{(m)}}V$ and if we choose $0<\varepsilon<\eta$ enough small then we may suppose that we have:

(i) $U(S)$ depends continuously on S in $N_m^+(T,\varepsilon)$,
(ii) $S_{U(S)H}^{(m)}$ is Fredholm and injective.

If we define $F(S)\varepsilon L(H^{(m)},H)$, $S\varepsilon N_m^+(T,\varepsilon)$ by the equation

$$F(S)(\bigoplus_{k=1}^{m}h_k)=S^{m-1}U(S)V(S)^*h_1+(I-V(S)V(S)^*)h_1+\sum_{k=2}^{m}S^{m-k}U(S)h_k$$

then obviously $F(S)$ is invertible and $F(S)^{-1}SF(S)$ has the matrix representation appearing in the statement with $S_{1,2}=V(S)$.

For every positive integer $m\ge 1$, we shall denote by Π_m, the canonical map of the integers onto the residue class field, mod m: $\{0,\dots,m-1\}$.

1.3. LEMMA. *Let* $T\in L(H)$ *be a nilpotent of order* $m\geq 2$ *of the form*

$$T=\begin{bmatrix} 0 & T_{1,2} & * \cdots\cdots & * \\ 0 & 0 & T_{2,3} \cdots & * \\ \vdots & & & \\ 0 & \cdots\cdots & \cdots\cdots & T_{m-1,m} \\ 0 & \cdots\cdots & \cdots\cdots & 0 \end{bmatrix}$$

where $T_{1,2}$ *is Fredholm, left-invertible, and* $T_{j,j+1}$, $j>1$ *is invertible. Let* $\{P_n\}_{n=1}^{\infty}\subset L(H)$ *be a sequence of finite-rank projections such that*

$$P_n \xrightarrow{s} I,\ P_n^* \xrightarrow{s} I,\ \|TP_n-P_nT\| \to 0\ .$$

Then we have

$$\Pi_m(\dim\ker T_{1,2}^*)=\lim_{n\to\infty}\Pi_m(\operatorname{rank} P_n)\ .$$

PROOF. Because the inner derivation Δ_T has closed range (see [1]), using an easy spectral argument, we can find a sequence $\{Q_n\}_{n=1}^{\infty}\subset L(H)$ of finite-rank projections commuting with T, such that $\|Q_n-P_n\|\to 0$. If Q_n has a matrix representation of the form

$$Q_n=\begin{bmatrix} Q_{1,1}^{(n)} & * & \cdots\cdots & * \\ 0 & Q_{2,2}^{(n)} & * \ \cdots & * \\ \vdots & & & \\ 0 & \cdots\cdots & \cdots\cdots & Q_{m,m}^{(n)} \end{bmatrix}$$

it is an easy exercise to show that we have $\operatorname{rank} Q_n=\sum_{k=1}^{m}\operatorname{rank} Q_{k,k}^{(n)}$.

Let $n_o\geq 1$ be such that

$$\operatorname{rank} Q_n=\operatorname{rank} P_n\ ,\qquad \ker Q_{1,1}^{(n)}\cap\ker T_{1,2}^*=\{0\}\ ,\qquad n\geq n_o\ .$$

Using the relations

$$Q_{1,1}^{(n)}T_{1,2}=T_{1,2}Q_{2,2}^{(n)}\ , \qquad Q_{k,k}^{(n)}T_{k,k+1}=T_{k,k+1}Q_{k+1,k+1}^{(n)}\ , \qquad k\geq 2$$

we derive

$$\text{rank } Q_{k,k}^{(n)}=\text{rank } Q_{m,m}^{(n)}\ , \qquad k\geq 2, \qquad \text{rank } Q_{1,1}^{(n)}=\text{rank } Q_{m,m}^{(n)}+\dim\ker T_{1,2}^{*}\ ,$$

if $n\geq n_o$. Thus we have

$$\text{rank } P_n=\text{rank } Q_n=\dim\ker T_{1,2}^{*}+m \text{ rank } Q_{m,m}^{(n)}\ , \qquad n\geq n_o$$

and this concludes the proof.

1.4. PROPOSITION. *For every* $m\geq 2$, $0\leq k,\ell\leq m-1$ *we have*

$$N_{m,k}^{+}(H)\cap N_{m,\ell}^{+}(H)=\emptyset\ , \qquad k\neq\ell\ .$$

PROOF. Suppose the contrary. Then there exist a compact operator K and an invertible operator A such that $q_1^{(k)}\oplus q_m^{(\infty)}=$ $=K+A^{-1}(q_1^{(\ell)}\oplus q_m^{(\infty)})A$. Let $\{P_n\}_{n=1}^{\infty}$ be a sequence of finite-rank, self-adjoint projections commuting with $q_1^{(k)}\oplus q_m^{(\infty)}$ and strongly convergent to the identity operator. Then applying Lemma 1.3 we derive the contradiction

$$k=\lim_{n\to\infty}\Pi_m(\text{rank } P_n)=\lim_{n\to\infty}\Pi_m(\text{rank } A^{-1}P_nA)=\ell$$

and this concludes the proof.

1.5. PROPOSITION. *For every* $m\geq 1$ *we have* $N_m^{+}(H)=\bigcup_{k=0}^{m-1}N_{m,k}^{+}(H)$. *Consequently* $N_m^{+}(H)$ *is invariant to similarities.*

PROOF. Let $T\varepsilon N_m^{+}(H)$ be given and let $F(T)$ be defined by Theorem 1.2. If $T'\varepsilon L(H^{(m)})$ is the matrix operator

$$T'=\begin{bmatrix} 0 & T_{1,2} & 0 & \cdots & \cdots & 0 \\ 0 & 0 & I & 0 & \cdots & 0 \\ \vdots & & & & & \\ 0 & \cdots & \cdots & \cdots & \cdots & I \\ 0 & \cdots & \cdots & \cdots & \cdots & 0 \end{bmatrix}$$

where $T_{1,2}$ is the (1,2)-entry of $F(T)^{-1}TF(T)$, then it is plain that T' is unitarily equivalent with a finite-rank perturbation of $q_1^{(k)} \oplus q_m^{(\infty)}$, $k=\Pi_m(\dim \ker T_{1,2}^*)$. But this means $T\varepsilon N_{m,k}^+(H)$ and the inclusion " $\subset$ " is proved. Conversely, if T-K is similar with $q_1^{(k)} \oplus q_m^{(\infty)}$ where K is compact, then T-K has a matrix representation of the form

$$T-K=\begin{bmatrix} 0 & T'_{1,2} & * & \cdots & * \\ 0 & 0 & T'_{2,3} & * \cdots & * \\ \vdots & & & & \\ 0 & \cdots & & & T'_{m-1,m} \\ 0 & \cdots & & & 0 \end{bmatrix}$$

with $T'_{j,j+1}$ a Fredholm operator, $1\le j\le m-1$. But this clearly implies that $T^{m-1}+T^*$ is Fredholm and $T\varepsilon N_m^+(H)$.

2. THE INDEX CONDITION

For every $T\varepsilon N_m^+(H)$ put

$$\operatorname{ind}_m T=\{k:0\le k\le m-1,\ T\varepsilon N_{m,k}^+(H)\}.$$

By Propositions 1.4 and 1.5 we derive that $\operatorname{ind}_m T$ is well defined.

2.1. THEOREM. *The function* ind_m *is continuous and for every* $T\varepsilon N_m^+(H)$, $K\varepsilon K(H)$, $A\varepsilon L(H)$, $0\notin\sigma(A)$ *we have*

$$\operatorname{ind}_m T=\operatorname{ind}_m A^{-1}TA=\operatorname{ind}_m(T+K)\ .$$

PROOF. We may suppose $m\ge 2$. Let $\varepsilon>0$, F be as in Theorem 1.2 and for every $S\varepsilon N_m^+(T,\varepsilon)$ let $S_{1,2}$ denote the (1,2)-entry of $F(S)^{-1}SF(S)$. Then as seen in the proof of Proposition 1.5 we have $S\varepsilon N_{m,k}^+(H)$, $k=\Pi_m(\dim \ker S_{1,2}^*)$, thus $\operatorname{ind}_m(S)=\Pi_m(\dim \ker S_{1,2}^*)$. But because $S_{1,2}$ is a continuous function of S in $N_m^+(T,\varepsilon)$, the continuity of ind_m follows. The rest of the proof follows from the definition of ind_m.

2.2. COROLLARY. *Let* $m\geq 1$, $0\leq k\leq m-1$ *be given. Then* $N^+_{m,k}(H)$ *is clopen in* $N^+_m(H)$.

PROOF. Using Propositions 1.4 and 1.5 it suffices to prove that $N^+_{m,k}$ is closed. But $N^+_{m,k}(H)$ will be closed because ind_m is continuous on $N^+_m(H)$ and we have

$$N^+_{m,k}(H)=\{T\in N^+_m(H):\operatorname{ind}_m T=k\}.$$

2.3. THEOREM. *Let* $T\in N^+_m(H)$ *be given. Then* ind_m *is constant on* $S^+(T)$.

PROOF. If $\operatorname{ind}_m T=k$ we have $S(T)\subset N^+_{m,k}(H)$ and by Corollary 2.2 we also have $S^+(T)\subset N^+_{m,k}(H)$. This implies $\operatorname{ind}_m S=k$, $S\in S^+(T)$.

3. THE RANGE CONDITION

For every $T\in L(H)$, let $R(T)$ denote the range of T and put

$$N^{+,\ell}_m(H)=\{T\in N^+_m(H):R(T^\ell)=R(T^\ell)^-\}.$$

3.1. LEMMA. *Let* $T\in N^+_m(H)$, $m\geq 2$ *be given and let* ε, F *be defined as in Theorem 1.2. Then for every* $S\in N^+_m(T,\varepsilon)$, $1\leq\ell\leq m$, *we have* $R(S^\ell)=R(S^\ell)^-$ *if and only if* $\operatorname{rank} S_{j,1}<\infty$, $m-\ell<j\leq m$.

PROOF. The range of S^ℓ is closed if and only if the range of $(F(S)^{-1}SF(S))^\ell$ is closed. A simple calculation shows that $(F(S)^{-1}SF(S))^\ell$ has a matrix representation of the form

$$(F(S)^{-1}SF(S))^\ell=\left[\begin{array}{cccccccc} * & \cdots\cdots & \overbrace{S_{1,1}S_{1,2} & S_{1,2}}^{\ell\ \text{columns}} & 0 & \cdots\cdots & & 0 \\ * & \cdots\cdots & S_{2,1}S_{1,2} & 0 & I & 0 & \cdots & 0 \\ \vdots & & & & & & & \\ \vdots & & & & & & & I \\ * & \cdots\cdots & S_{m,1}S_{1,2} & 0 & \cdots\cdots & & & 0 \end{array}\right]\begin{array}{l}\left.\begin{array}{l} \\ \\ \\ \\ \end{array}\right\} m-\ell\ \text{rows} \\ \\ \end{array}$$

Suppose that $(F(S)^{-1}SF(S))^{\ell}$ has closed range. Because all *-entries are compact, $S_{1,2}$ is a Fredholm isometry and $S_{j,1}S_{1,2}$, $1\le j\le m$ is compact we can derive that $S_{j,1}S_{1,2}$ as well as $S_{j,1}$ is finite--rank, $m-\ell<j\le m$. The reverse implication follows analogously.

3.2. PROPOSITION. *For every* $m\ge 1$ *we have*

$$N_m^{+,1}(H)\supset \dots \supset N_m^{+,m}(H).$$

PROOF. It follows from Lemma 3.1.

3.3. THEOREM. *For every* $M\in N_m^{+,\ell}(H)$, $m\ge 1, 1\le \ell\le m$, *we have*

$$S^{+}(M)\subset N_m^{+,\ell}(H) .$$

PROOF. We shall assume $m\ge 2$. Let $T\in S_m^{+}(H)$ be given and let $\varepsilon>0$, F be as in Theorem 1.2. Without loss of generality we may suppose that we have $M\in N_m^{+}(T,\varepsilon)$ and $\dim\ker S_{1,2}^{*}$=constant for $S\in N_m^{+}(T,\varepsilon)$. Let P_j, $1\le j\le m$ denote the orthogonal projection of $H^{(m)}$ onto the j copy of H. Suppose $S\in N_m^{+}(T,\varepsilon)\cap S(M)$ and let $A=\{A_{i,j}\}$ be an invertible operator A such that

$$F(M)^{-1}MF(M)A=AF(S)^{-1}SF(S) .$$

Since by Lemma 3.1 we have rank $M_{m,1}<\infty$ and

$$\operatorname{rank} P_mAF(S)^{-1}SF(S)=\operatorname{rank} P_mF(M)^{-1}MF(M)A\le \operatorname{rank} M_{m,1}$$

$$P_mAF(S)^{-1}SF(S)=\begin{bmatrix} 0 & 0 & \dots & 0 \\ \vdots & & & \\ 0 & * & \dots & 0 \\ * & A_{m,1}S_{1,2} & A_{m,2}\ \dots & A_{m,m-1} \end{bmatrix}$$

we derive rank $P_mA(I-P_m)\le a_m<\infty$ where a_m depends on M and T only. Proceeding analogously we derive in a finite number of steps

$$\operatorname{rank} P_jA(I-\sum_{k\ge j} P_k)\le a_j<\infty , \qquad m-\ell<j\le m$$

where a_j depends on M and T only. Now using again the equation

$$F(M)^{-1}MF(M)A=AF(S)^{-1}SF(S)$$

we derive first $A_{i,j}\varepsilon K(H)$, i>j and then

$$M_{1,2}A_{2,2}-A_{1,1}S_{1,2}\varepsilon K(H),\ A_{2,2}-A_{j,j}\varepsilon K(H),\quad j\geq 2\ .$$

Since A is Fredholm of index 0, denoting by A′ the upper triangular part of A, we have

$$0=\text{ind}A'=\sum_{j=1}^{m}\text{ind}A_{j,j}=\text{ind}A_{1,1}+(m-1)\text{ind}A_{2,2}$$

$$\text{ind}M_{1,2}A_{2,2}=\text{ind}M_{1,2}+\text{ind}A_{2,2}=\text{ind}A_{1,1}S_{1,2}=\text{ind}A_{1,1}+\text{ind}S_{1,2}$$

and this implies $\text{ind}A_{j,j}=0$, $1\leq j\leq m$. But we also have

$$\dim\ker A_{j,j}\leq b_j<\infty\ ,\qquad m-\ell<j\leq m\ ,$$

where b_j depends on M and T only. Indeed, because A is surjective and rank $P_mA(I-P_m)\leq a_m$ we have $\dim\ker A^*_{m,m}=\dim\ker A_{m,m}\leq a_m$ and we can take $b_m=a_m$. Analogously, we derive $\dim\ker A_{j,j}<b_j$, $m-\ell<j\leq m$. Define $B\varepsilon L(H^{(m)})$ by the matrix

$$B_{i,j}=A_{i,j}\ ,\qquad i\leq m-\ell \text{ or } i>m-\ell,\ j\leq i$$
$$B_{i,j}=0\ ,\qquad i>m-\ell,\ j<i.$$

Since $\text{rank}(A-B)\leq a<\infty$, where a depends on M and T only and we have

$$\text{rank}(BF(S)^{-1}SF(S)-F(M)^{-1}MF(M)B)\leq 2a$$

we derive

$$\text{rank}(A_{m,m}S_{m,1}-M_{m,1}A_{1,1})\leq 2a$$
$$\text{rank}(A_{m-1,m-1}S_{m-1,1}+A_{m-1,m}S_{m,1}-M_{m-1,1}A_{1,1})\leq 2a$$
$$\cdots\cdots\cdots\cdots\cdots\cdots\cdots\cdots\ .$$

But this implies rank $S_{j,1} \leq c_j$, $m-\ell<j\leq m$, where c_j depends on M and T only and because $T_{j,1}=\lim S_{j,1}$ we infer that $F(T)^{-1}T^{\ell}F(T)$ has closed range or equivalently $T\in N_m^{+,\ell}(H)$.

REFERENCES

1. Apostol, C.: Inner derivations with closed range, *Revue Roumaine Math.Pures Appl.* 21(1976), 249-265.
2. Apostol, C.; Stampfli, J.G.: On derivation ranges, *Indiana Univ.Math.J.* 25(1976), 857-869.
3. Barria, J.; Herrero, D.A.: Closure of similarity orbits of nilpotent operators. I. Finite-rank operators, *J.Operator Theory* 1(1979), 177-185.
4. Fillmore, P.A.; Stampfli, J.G.; Williams, J.P.: On the essential spectrum, the essential numerical range and a problem of Halmos, *Acta Sci.Math.* (Szeged) 33(1972), 179--192.
5. Herrero, D.A.: Clausura de las órbitas de similaridad de operadores en espacios de Hilbert, *Rev.Un.Mat.Argentina* 27(1977), 244-260.
6. Herrero, D.A.: *Approximation of Hilbert Space Operators. I*, Pitman, to appear.
7. Kato, T.: *Perturbation Theory for Linear Operators*, Springer--Verlag, 1966.

C.Apostol
Department of Mathematics,
INCREST,
Bdul Păcii 220, 79622 Bucharest,
Romania.

INVARIANT SUBSPACES AND FUNCTIONAL REPRESENTATIONS FOR THE $C_{1\cdot}$ CONTRACTIONS

B.Beauzamy

Let E be an infinite-dimensional, separable, Banach space. An operator T, from E into itself, is called a $C_{1\cdot}$ contraction if $||T||=1$ and, for all $x\neq 0$, $T^n x \not\to 0$, $(n \to +\infty)$. (This terminology is due to Sz.-Nagy-Foiaş [8].)

The first result concerning the existence of invariant subspaces for a $C_{1\cdot}$ contraction is a special case of a Theorem of J.Wermer [9]: if T is invertible, and if $\sum\limits_{k\geq 0} \frac{\log||T^{-k}||}{1+k^2}<+\infty$, then T has nontrivial hyperinvariant subspaces.

Then, in 1965, I.Colojoară and C.Foias proved [6] that if T is defined on a reflexive space, and is C_{11}, that is moreover ${}^tT^n x \not\to 0$, $(n \to +\infty)$, then T has nontrivial hyperinvariant subspaces.

In 1979, the author obtained the following result [1]:

THEOREM 1. *Let* E *be a separable Banach space,* T *an invertible* $C_{1\cdot}$ *contraction. We assume that there is a point* $y\neq 0$ *and a sequence* $(\rho_n)_{n\geq 0}$ *of real numbers, satisfying*

$$(1)\qquad \begin{cases} \rho_{m+n}\leq \rho_m\cdot\rho_n\,, \quad \rho_n\geq 1 \text{ for all } n, m, \\ \text{and} \\ \sum\limits_{n\geq 0} \dfrac{\log \rho_n}{1+n^2}<+\infty, \end{cases}$$

such that $||T^{-n}y||\leq \rho_n$ *for all* n. *Then* T *has nontrivial hyperinvariant subspaces.*

Another theorem of this type was also proved in [1] in the noninvertible case.

Theorem 1 relies upon the following construction: if $f\in A(\mathbb{T})$, that is, $f=\sum_{k\in\mathbb{Z}} c_k e^{ik\theta}$ with $\Sigma|c_k|<+\infty$, we put $\Psi_m(f)=\sum_{k\geq -m} c_k T^{k+m}$, $m\geq 0$. We say that a subspace F given by $F=\{z\in E,\ \Psi_m(f)z \to 0,\ (m\to+\infty)\}$, for some $f\in A(\mathbb{T})$, is of functional type. If, for all $z\in E$, $\Psi_m(f)z \to 0$, $(m\to+\infty)$, we say that T satisfies an asymptotic functional equation of function f.

The invariant subspaces obtained in [1] are of functional type, and we prove in [1] that the conditions (1) are the best possible in order to obtain this type of invariant subspace. This proves that this method cannot apply, for example, to an invertible operator T satisfying

$$||T^{-n}x||\geq(1+\varepsilon)^n C(x),\qquad C(x)>0,\text{ for all } n\geq 0. \tag{2}$$

A typical example of such an operator is the weighted shift, on $\ell^2(\mathbb{Z})$, defined by $Te_n=w_n e_{n+1}$, with $w_n=1/4$ for $n<0$, and $w_n=1$ for $n\geq 0$. This operator was studied by the author in [3], where is proved that for every $x\in\ell^2(\mathbb{Z})$,

$$x\notin\overline{\text{span}}\{Tx\ ,\ T^2x,\ldots\}\ . \tag{3}$$

So this operator has indeed many invariant subspaces.

An attempt to find a formalization of this type of operator was made by the author in [2], dealing only with Hilbert space. The following result was obtained:

THEOREM 2. *For a* $C_{1\cdot}$ *contraction* T *on a Hilbert space, one of the following is true:*

a) *For every* $x,y\in E$, *if* $\lambda_j(x,y) = \lim\limits_{m\to+\infty}\langle T^{m+j}x,\ T^m y\rangle$, $(j\in\mathbb{Z})$, *then the sequence* $(\lambda_j(x,y))_{j\in\mathbb{Z}}$ *is a cyclic vector for the usual shift in* $c_o(\mathbb{Z})$;

b) T *satisfies an asymptotic functional equation;*

c) T *has a nontrivial invariant subspace.*

We shall come back later on this theorem, and obtain further precisions. If E is a Banach space, we obtained in [4] the following result for operators satisfying (2):

THEOREM 3. *If* T *satisfies* (2) *then either* T *has an eigenvalue or the iterates* $(T^{-n}x)_{n\geq 0}$ *are* ω*-linearly independent, that is, if a series* $\sum_{k\geq 0} \alpha_k T^{-k}x=0$, *then* $\alpha_k=0$ *for all* $k\geq 0$.

This result is obviously weaker than (3), but is of the same nature.

In [7], M.Rome established the links between the minimal unitary dilation of T (in the sense of Sz.-Nagy-Foiaş [8]), or, more precisely, the $*$-residual part of it, and the tools used in [1] by the author. It follows from his results that if T satisfies an asymptotic functional equation of function f $(f\in A(\mathbb{T}))$, then f must vanish identically on the spectrum of the $*$-residual part of the minimal unitary dilation, that is, on a set of positive measure of $\mathbb{T}$.

Later, A.Atzmon showed that this $*$-residual part can be simply defined as follows (E being a Hilbert space):

For $x\in E$, put $[x]=\lim_{m\to+\infty} \|T^m x\|$, and $[x,y]=\lim_{m\to+\infty} \langle T^m x, T^m y\rangle$. This is a scalar product. Let H be the completion of E for the norm $[\cdot]$, and $\tilde{T}$ the extension of T to H. Then $\tilde{T}$ is the required operator. It is a surjective isometry on H, if T has dense range.

The results which follow now were established by M.Rome and the author [5]. T is assumed to be invertible, $C_{1\cdot}$, and defined on a Hilbert space.

PROPOSITION 4. *For every* $x,y\in H$, *the Fourier series* $\sum_{k\in\mathbb{Z}} [\tilde{T}^k x,y]e^{ik\theta}$ *defines a function* $\Lambda_{x,y}(\theta)$, *which is Lebesgue-integrable.*

For x=y, we just put Λ_x instead of $\Lambda_{x,x}$. Then, the following proposition is a formalization of Property (3).

PROPOSITION 5. *For every* $x\in E$,

$$\exp \int \log \Lambda_x(\theta)\frac{d\theta}{2\pi} = \operatorname*{Inf}_{m\geq 0} \operatorname{dist}(T^m x, \overline{\operatorname{span}}\{T^{m+1}x,\ldots,T^{m+k}x,\ldots\})^2 =$$

$$= \operatorname*{Inf}_{m\geq 0} \operatorname{dist}(T^m x, \overline{\operatorname{span}}\{T^{m-1}x,\ldots,T^{m-k}x,\ldots\})^2.$$

So, if $\log \Lambda_x$ is integrable, $T^m x \notin \overline{\mathrm{span}}\{T^{m+1}x, T^{m+2}x, \ldots\}$. This requires that the spectrum of $\tilde{T}$ is the whole unit circle, because out of this spectrum, all the functions $\Lambda_x(-\theta)$, $(x \in E)$ vanish. The spectrum of $\tilde{T}$ is contained in $\sigma(T) \cap \{|z|=1\}$.

The functions Λ_{x_o} can be used to obtain a functional representation of the series using the iterates $(T^k x_o)_{k\in\mathbb{Z}}$ of a point $x_o \in E$. We put $\check{\Lambda}_x(\theta) = \Lambda_x(-\theta)$.

PROPOSITION 6. *For every convergent series* $\sum_{k\in\mathbb{Z}} a_k T^k x_o$, *the Fourier series* $\sum_{k\in\mathbb{Z}} a_k e^{ik\theta}$ *defines a function* φ *of* $L^2(\check{\Lambda}_{x_o}(\theta)\cdot\frac{d\theta}{2\pi})$: *the sums* $\sum_{-N}^{M} a_k e^{ik\theta}$ *converge in this space when* $M,N \to +\infty$; *moreover, one has* $\check{\Lambda}_x = |\varphi|^2 \check{\Lambda}_{x_o}$, *if* $x = \sum_{k\in\mathbb{Z}} a_k T^k x_o$.

A similar result holds if the series $\sum_{k\in\mathbb{Z}} a_k T^k x_o$ converges only in the Abel sense, that is, the limit of $\sum_{k\in\mathbb{Z}} a_k r^{|k|} T^k x_o$ exists when $r \to 1^-$. If the series $\sum_{k\in\mathbb{Z}} a_k T^k x_o$ converges unconditionally, then $\varphi \in L^2$.

From these results, we deduce the following proposition, which extends Theorem 3 and a result of Sz.-Nagy-Foiaş [8]:

THEOREM 7. *If* T *satisfies* (2), *the only series* $\sum_{k\in\mathbb{Z}} a_k T^k x_o$ *converging to* 0 *in the Abel sense, such that the corresponding function* φ *belongs to the Nevanlinna class (that is, satisfies* $\sup_{0<r<1} \int \log^+ |\varphi(re^{i\theta})| d\theta < +\infty$) *is the series with all* a_k*'s=0.*

REFERENCES

1. Beauzamy, B.: Sous-espaces invariants de type fonctionnel dans les espaces de Banach, *Acta Math.* 144 (1980), 65-82.

2. Beauzamy, B.: Sous-espaces invariants pour les contractions de classe $C_{1\cdot}$, et vecteurs cycliques pour le shift de $c_o(\mathbb{Z})$, *J. Operator Theory* 7 (1982), 125-137.

3. Beauzamy, B.: A weighted shift with no cyclic vector, *J.*

Operator Theory 4(1980), 287-288.

4. Beauzamy, B.: Une propriété de régularité pour les itérés inverses des contractions de classe $C_{1.}$, *C.R.Acad.Sci.Paris Sér.A* 290(1980), 467-469.

5. Beauzamy, B.; Rome, M.: Représentation fonctionnelle des séries convergentes utilisant les itérés d'un point par une contraction de classe $C_{1.}$, *C.R.Acad.Sci.Paris Sér.A* 292 (1981), 963-965.

6. Colojoară, I.; Foiaş, C.: *Theory of generalized spectral operators*, Gordon and Breach, 1968.

7. Rome, M.: Dilatations isométriques d'opérateurs et sous--espaces invariants, *J.Operator Theory* 6(1981), 39-50.

8. Sz.-Nagy, B.; Foiaş, C.: *Analyse harmonique des opérateurs dans l'espace de Hilbert*, Akademiai Kiado-Budapest, 1966.

9. Wermer, J.: The existence of invariant subspaces, *Duke Math. J.* 19(1952), 615-622.

B.Beauzamy
Département de Mathématiques,
Université de Lyon 1,
43, bd. du 11 Novembre 1918,
69621 Villeurbanne Cedex,
France.

INTERTWININGS AND HYPERINVARIANT SUBSPACES

B.Chevreau

Let E be a Banach space, and let $L(E)$ denote the Banach algebra of all bounded linear operators on E. A *nontrivial hyperinvariant subspace* for an operator A in $L(E)$ is a nonzero, proper, (closed) subspace of E which is invariant under any operator in $\{A\}'$, the commutant of A.

The purpose of this note is to present a general principle which leads to the existence of nontrivial hyperinvariant subspaces for certain operators. In particular we recapture many of the results asserting the existence of hyperinvariant subspaces for operators satisfying growth conditions either global or local ([3], [4], [6], [7], [10], [12]).

Our principle is based on the construction of intertwinings between a given operator and an operator of multiplication by z on a certain Banach algebra of functions. We already used it in a special case to make certain reductions to "good cases" in applications of the S.Brown technique [2]. An additional reason which motivates this approach is its close connection with local spectral theory.

The paper is organized as follows. In Section 1 we present the general principle which boils down to the well-known fact that if two operators have disjoint spectra then these two operators cannot be nontrivially intertwined. Section 2 contains some preliminary material on regular Banach algebras and intertwinings. The main results are given in Section 3. In Section 4 we show the connection of our approach with local spectral theory. Section 5 contains the applications to operators satisfying growth conditions. In Section 6 we recall how this principle was used in connection with the S.Brown technique.

1. THE GENERAL PRINCIPLE

We begin by recalling the following well-known fact.

PROPOSITION 1.1. ([9]). *Let E and F be two Banach spaces, let* A *and* B *be operators on E and F respectively, and let* $T:E\to F$ *be a bounded linear operator intertwining* B *with* A (i.e. $AT=TB$). *If the spectra of* A *and* B *are disjoint then* $T=0$.

The following proposition contains the key idea of our way of connecting intertwinings and hyperinvariant subspaces.

PROPOSITION 1.2. *Let E be a Banach space, let* $A\in L(E)$, *let* $U_i\in L(B_i)$ $i=1,2$ *where B_1 and B_2 are Banach spaces, and let* $T_1\in L(B_1,E)$, $T_2\in L(B_2,E^*)$ *such that:*

1. $AT_1=T_1U_1$,
2. $A^*T_2=T_2U_2$,
3. $\sigma(U_1)\cap\sigma(U_2)=\emptyset$.

Then A *has a nontrivial hyperinvariant subspace.*

PROOF. (The reader is advised to follow the intertwinings with the help of the usual commutative diagrams.) We will denote by j the canonical embedding of E into its bidual E^{**}. Recall that $A^{**}j=jA$ and that $j(A)$ is weak* dense in A^{**}. Let B be any operator commuting with A. Then $A(BT_1)=(BT_1)U_1$. This equality combined with the intertwinings $A^{**}j=jA$ and $U_2^*T_2^*=T_2^*A^{**}$ (this latter one being the dual of (2)) yields the following equality

$$U_2^*(T_2^*jBT_1)=(T_2^*jBT_1)U_1 .$$

Since $\sigma(U_1)\cap\sigma(U_2^*)=\emptyset$ we deduce from Proposition 1.1 that $T_2^*jBT_1=0$, that is $\mathrm{Ran}(BT_1)\subset\mathrm{Ker}(T_2^*j)$ for every B in $\{A\}'$. Since T_2^* is continuous when E^{**} and B_2^* are equipped with their respective weak* topologies the operator T_2^*j is nonzero (otherwise the weak* density of $j(E)$ in E^{**} would imply $T_2^*=0$ and hence $T_2=0$).

Thus $\mathrm{Ker}(T_2^*j)$ is a proper subspace of E and the closed linear span of $\mathrm{Ran}(BT_1)$ when B runs over $\{A\}'$ is a nontrivial hyperinvariant subspace for A.

REMARK. It is easily seen that the setting of Proposition 1.2 is compatible with quasisimilarity in the following sense. Suppose A is an operator in $L(E)$ for which there exist operators U_1 and U_2 nontrivially intertwinable with A and A^* respectively and suppose that B is quasisimilar to A; then U_1 and U_2 are also nontrivially intertwinable with B and B^* respectively.

In order to obtain concrete applications of Proposition 1.2 we will specialize it to the case where $\mathcal{B}_1$ and $\mathcal{B}_2$ are Banach algebras of functions, and U_1 , U_2 operators of multiplication by z on these algebras. In the next section we present some preliminary material on the type of algebras we are going to use, as well as some notions and results relevant to the context of the existence of an intertwining of multiplication by z with a given operator A.

2. PRELIMINARIES

Throughout this paper a (commutative, unital) Banach algebra $\mathcal{B}$ will be said to satisfy Condition (*) if it satisfies the following four requirements:

1. $\mathcal{B}$ is semi-simple (i.e. the Gelfand transform is one-to-one); this fact enables us to view elements of $\mathcal{B}$ as (continuous) functions on the maximal ideal space $\mathrm{Max}(\mathcal{B})$, of $\mathcal{B}$.
2. $\mathrm{Max}(\mathcal{B})$ is a (compact) subset of $\mathbb{C}$.
3. The position function $(z \to z)$ on $\mathrm{Max}(\mathcal{B})$ belongs to $\mathcal{B}$.
4. $\mathcal{B}$ is regular; that is given a closed subset E of $\mathrm{Max}(\mathcal{B})$ and $\lambda \in \mathrm{Max}(\mathcal{B}) \setminus E$ there exists f in B such that $f|E=0$ and $f(\lambda)=1$.

For details on regular semisimple Banach algebras the reader is referred to [8, Chapter VIII]. We recall from there that such algebras are in fact normal (i.e. given any two disjoint closed subsets E and F of $\mathrm{Max}(\mathcal{B})$ there exists f in $\mathcal{B}$ such that $f|E=0$ and $f|F=1$). For an ideal I in $\mathcal{B}$ we denote by h(I) the (closed) subset of $\mathrm{Max}(\mathcal{B})$ defined by

$$h(I)=\{x \in \mathrm{Max}(\mathcal{B}) \mid \forall f \in I,\ f(x)=0\}.$$

It is well-known (and easily checked) that $\mathrm{Max}(\mathcal{B}/I)=h(I)$ (but $\mathcal{B}/I$ is not necessarily semisimple). For a closed subset E of $\mathrm{Max}(\mathcal{B})$ let $k(E)$ denote the ideal of all functions in $\mathcal{B}$ vanishing on E. Then the regularity of $\mathcal{B}$ implies that $h(k(E))=E$ [8, Chapter VIII]. One sees easily that $\mathcal{B}/k(E)$ satisfies (*) whenever $\mathcal{B}$ does.

Suppose that $\mathcal{B}$ satisfies (*) and let T be a bounded linear map from $\mathcal{B}$ into E. For f in $\mathcal{B}$ we define $fT:\mathcal{B}\to E$ by $(fT)(g)=T(fg)$ (in other words, $fT=TM_f$ where M_f denotes the operator of multiplication by f in $\mathcal{B}$) and denote by I_T the annihilator of T, i.e. $I_T=\{f\in\mathcal{B}:fT=0\}$. The support of T (notation $\Sigma(T)$) is the complement (in $\mathrm{Max}(\mathcal{B})$) of the largest open subset Ω of $\mathrm{Max}(\mathcal{B})$ such that $T(g)=$ $=0$ whenever $\mathrm{supp}\, g\subset\Omega$. It turns out that $\Sigma(T)=h(I_T)$ (this was proved in [3] for bounded linear functionals and in the case of a Beurling algebra, but the proof works for any algebra $\mathcal{B}$ which satisfies (*) and any E-valued mapping T defined on $\mathcal{B}$). Clearly we have $\Sigma(fT)\subset\mathrm{supp}\, f\cap\Sigma(T)$ for any f in $\mathcal{B}$.

Suppose now that T intertwines U (the operator of multiplication by z on $\mathcal{B}$) with $A\in L(E)$. Since $U(I_T)\subset I_T$, U induces a map $\hat{U}:\mathcal{B}/I_T\to\mathcal{B}/I_T$. Note also that $I_T\subset\mathrm{Ker}\, T$; thus T can be factored out $T=\hat{T}\circ\pi$ where π is the quotient map $\mathcal{B}\to\mathcal{B}/I_T$. It is easy to see that $\hat{T}\hat{U}=A\hat{T}$. This "induced" intertwining will sometimes be helpful. Unfortunately as mentioned before $\mathcal{B}/I_T$ need not satisfy (*). But the above considerations apply as well if, instead of I_T, we use any ideal I contained in Ker T, for instance $I=k(E)$ where E is any closed set containing $\Sigma(T)$ in its interior (so that any f vanishing on E has its support disjoint from $\Sigma(T)$; therefore we have $fT=0$ and, in particular, f belongs to Ker T). With $I=k(E)$ we have $h(I)=E$ and $\mathcal{B}/k(E)$ is an algebra satisfying Condition (*), whose elements can be identified with functions on E. The above quotient map is now the restriction map ($f\overset{\pi}{\to}f|E$) and $\hat{U}$ is multiplication by z on $\mathcal{B}/k(E)$. In particular we have $\sigma(\hat{U})=E$. Clearly in view of Proposition 1.2 it is desirable to obtain operators U with small spectrum. This is achieved by means of the following proposition.

PROPOSITION 2.1. *Suppose that* $\mathcal{B}$ *satisfies* (*) *and that* T *is a nonzero operator intertwining* U *with* A. *Let* λ_o *belong to* $\Sigma(T)$

and let $\varepsilon>0$ *be given. Then there exists an algebra* $\mathcal{B}'$ *satisfying* (*)*, and a nonzero operator* T' *intertwining* U' *with* A *such that*

$$\sigma(U') \subset \{\lambda\in\mathbb{C}: |\lambda-\lambda_o|\le\varepsilon\}.$$

PROOF. Let f be a function in $\mathcal{B}$ such that f is equal to one in a neighborhood of λ_o, and $\operatorname{supp} f \subset \{\lambda\in\mathbb{C}: |\lambda-\lambda_o|<\varepsilon/2\}$. The operator $T_1=fT$ is nonzero (otherwise $T=(1-f)T$ and $\lambda_o\notin\Sigma(T)$) and intertwines U with A. The set $E=\{\lambda\in\operatorname{Max}(\mathcal{B}): |\lambda-\lambda_o|\le\varepsilon\}$ contains $\Sigma(T_1)$ in its interior. Applying the considerations preceding Proposition 2.1 to the intertwining T_1 and the set E we obtain the desired result (with $\mathcal{B}'=\mathcal{B}/k(E)$, $U'=U$, and $T'=T_1$).

3. MAIN RESULTS

In this section we give the two main theorems.

THEOREM 3.1. *Let* A *in* $L(E)$ *and suppose that we can find algebras* $\mathcal{B}_1$ *and* $\mathcal{B}_2$ *satisfying Condition* (*) *and nonzero operators* $T_1\in L(\mathcal{B}_1,E)$, $T_2\in L(\mathcal{B}_2,E^*)$ *such that* (U_i, $i=1,2$, *denotes the operator of multiplication by* z *on* $\mathcal{B}_i$)

(1) $AT_1=T_1U_1$,

(2) $A^*T_2=T_2U_2$,

(3) $\Sigma(T_1)\cup\Sigma(T_2)$ *contains more than one point.*

Then A *has a nontrivial hyperinvariant subspace.*

PROOF. Choose λ_1 in $\Sigma(T_1)$ and λ_2 in $\Sigma(T_2)$ such that $\lambda_1\neq\lambda_2$. Applying Proposition 2.1 to the intertwinings (1) and (2) with respect to λ_1 and λ_2 and with $\varepsilon<|\lambda_1-\lambda_2|/2$ we obtain intertwinings to which Proposition 1.2 applies. The result follows.

Since Hypothesis (3), cannot always be formulated in a natural way it would be desirable to be able to delete it. The next result shows that this is possible under some additional condition on the ideals of one of the algebras $\mathcal{B}_1$, $\mathcal{B}_2$.

THEOREM 3.2. *Let* A, $\mathcal{B}_1$, $\mathcal{B}_2$, T_1, T_2 *such that Hypotheses* (1) *and* (2) *of Theorem* 3.1 *are satisfied. Suppose that one of the algebras* $\mathcal{B}_1$, $\mathcal{B}_2$ *has the property that whenever* h(I) *is a singleton the ideal* I *is finite codimensional. Then either* A *is a multiple of the identity or* A *has a nontrivial hyperinvariant subspace.*

PROOF. Say $\mathcal{B}_1$ has the mentioned property. We just have to deal with the case $\Sigma(T_1)=\{\lambda_1\}$. As noticed in the discussion preceding Proposition 2.1 we have a nontrivial intertwining between the induced operator $\hat{U}$ acting on $\mathcal{B}_1/I_T$ and A. Since $h(I_T)=\{\lambda_1\}$, $\mathcal{B}_1/I_T$ is finite dimensional and there exists a nonzero polynomial p such that $p(\hat{U})=0$. We have $p(A)\hat{T}=Tp(\hat{U})=0$ and Ker p(A) is a nonzero hyperinvariant subspace for A (it contains Ran T). If Ker $p(A)\neq E$ we are done. If Ker $p(A)=E$ we can as well assume that p is the minimal polynomial satisfying this condition. If p is reducible $p=p_1p_2$, deg $p_i\geq 1$, $i=1,2$, then Ker $p_1(A)$ is a nontrivial hyperinvariant subspace for A. If p is irreducible then A is a multiple of the identity.

4. CONNECTION WITH LOCAL SPECTRAL THEORY

Recall that an operator $A\in L(E)$ is said to have the single valued extension property (in short SVEP) if whenever f is an E-valued analytic function on an open set Ω of $\mathbb{C}$ such that $(A-\lambda)f(\lambda)=0$ for $\lambda\in\Omega$ then f=0. If A has the SVEP, then for any x in E the function $\lambda\to(A-\lambda)^{-1}x$ has a unique maximal analytic extension $R_A^x(\lambda)$ whose domain of definition is called the local resolvent set of x with respect to A (notation $\rho_A(x)$). The local spectrum of x with respect to A is by definition $\sigma_A(x)=\mathbb{C}\setminus\rho_A(x)$. For details on local spectral theory see [1] or [6].

Suppose that $\mathcal{B}$ is an algebra satisfying Condition (*) and let T be a nonzero intertwining of U (multiplication by z on B) with A in $L(E)$. The following proposition shows that local spectral theory is relevant to that setting.

PROPOSITION 4.1. *Let* A *be an operator satisfying the* SVEP *and let* T *be a nonzero intertwining of* U, *the operator of multiplication by* z *on an algebra* $\mathcal{B}$ *satisfying* (*), *with* A. *Let* x=T(1). *Then* $\sigma_A(x)\subset\Sigma(T)$.

PROOF. Let E be any closed set having $\Sigma(T)$ in its interior (relative to $\mathrm{Max}(\mathcal{B})$). We know (cf. considerations preceding Proposition 2.1) that $T=\hat{T}\circ\pi$ where π is the canonical map $\mathcal{B}\to\mathcal{B}'=\mathcal{B}/k(E)$ and that $A\hat{T}=\hat{T}\hat{U}$ where $\hat{U}$ is multiplication by z on $\mathcal{B}'$. The identity function $(z\to z)$ on E is an element of $\mathcal{B}'$ whose spectrum is E; thus its resolvent $\lambda\to R_\lambda$ is analytic on $\mathbb{C}\setminus E$. Thus the function $G(\lambda)=$ $=T\circ R_\lambda$ is an E-valued analytic function on $\mathbb{C}\setminus E$. From $(z-\lambda)R_\lambda(z)=1$ and $\hat{T}\hat{U}=A\hat{T}$ we obtain easily $(A-\lambda)G(\lambda)=\hat{T}(1)$; but $\hat{T}(1)=T(1)=x$ thus $G(\lambda)$ coincides with the local resolvent $R_A^x(\lambda)$ and $\sigma_A(x)\subset E$. Since $\bigcap\{E:\overset{\circ}{E}\supset\Sigma(T)\}=\Sigma(T)$ we obtain $\sigma_A(x)\subset\Sigma(T)$.

For operators having the SVEP there is a local functional calculus at any vector x based on the local resolvent (in the way the Riesz-Dunford functional calculus is based on the (global) resolvent). We briefly recall its definition (see for example [11, Theorem 2.2] where this local functional calculus is defined in a more general setting).

Let A in $L(E)$ having the SVEP and let $x\in E$. For any function f analytic in a neighborhood Ω of $\sigma_A(x)$ choose a system of curves Γ surrounding $\sigma_A(x)$ in Ω in the sense of [11] and let

$$S_A^x(f)=\frac{1}{2\pi i}\int_\Gamma f(\xi)R_A^x(\xi)d\xi \ .$$

(One shows easily that S(f) does not depend on Γ.) We sum up the essential features of this local functional calculus in the following proposition. (Here $A(X)$ denotes the algebra of functions analytic in a neighborhood of X and as usual, R(X) denotes the closure of the algebra of rational functions with poles off X in the algebra C(X), of continuous functions on X equipped with the sup norm.)

PROPOSITION 4.2. *Let* A *be an operator having the* SVEP. *Then the map* $S_A^x:A(\sigma_A(x))\to E$ *satisfies the following properties*:

(1) $S_x^A(1)=x$ *and* $S_x^A(z\to\frac{1}{z-\lambda})=R_x^A(\lambda)$ $\lambda\notin\sigma_A(x)$;

(2) $S_x^A(zf)=AS_x^A(f)$;

(3) *Let* X *be a compact subset of* $\mathbb{C}$ *containing* $\sigma_A(x)$ *in its interior. There exists a constant* M_X *such that for any* f *in* R(X), $\|S_x^A(f)\| \le M_X \|f\|$.

(4) S_x^A *is uniquely determined by* (1).

We conclude this section by showing that a well-known result of local spectral theory fits into the framework of Proposition 1.2.

PROPOSITION [1]. *Let* A *in* $L(E)$ *and suppose that* A *and* A^* *have the* SVEP. *Suppose moreover that there exist nonzero vectors* x *in* L, y *in* E^* *such that* $\sigma_A(x) \cap \sigma_{A^*}(y) = \emptyset$. *Then* A *has a nontrivial hyperinvariant subspace.*

PROOF. Choose compact subsets X_1 and X_2 in $\mathbb{C}$ such that $X_1 \cap X_2 = \emptyset$, $\sigma_A(x) \subset \overset{\circ}{X}_1$ and $\sigma_{A^*}(y) \subset \overset{\circ}{X}_2$. Let $B_1 = R(X_1)$, $B_2 = R(X_2)$, $T_1 = S_x^A | R(X_1)$, $T_2 = S_y^{A^*} | R(X_2)$, and U_1 and U_2 the operators of multiplication by z on B_1 and B_2. By (3) and (2) of Proposition 4.1 T_1 and T_2 are (nonzero) intertwinings of A and A^* with respectively U_1 and U_2. Since $\sigma(U_1) = X_1$ and $\sigma(U_2) = X_2$ we conclude from Proposition 1.2 that A has a nontrivial hyperinvariant subspace.

5. OPERATORS SATISFYING GROWTH CONDITIONS

In this section we show how a recent result of Atzmon [3] can be proved by the above method. Following [3] we say that a sequence $\rho = (\rho_n)_{n\in\mathbb{Z}}$ is a Beurling sequence if $\rho_n \ge 1$, $n\in\mathbb{Z}$, $\rho_{n+m} \le \rho_n \rho_m$, $n,m\in\mathbb{Z}$, and $\sum_{n\in\mathbb{Z}} (\log \rho_n)/(1+n^2) \le \infty$. The associated Beurling algebra is the Banach algebra A_ρ that consists of those continuous functions f on the unit circle $\mathbb{T}$ whose sequence of Fourier coefficients $\{\hat{f}(n)\}$ belongs to $\ell^1(\rho)$ (the norm on A_ρ being $\|f\|_\rho = \sum_{n\in\mathbb{Z}} |\hat{f}(n)|\rho_n$). From [8, Chap. VIII] it follows that $\mathrm{Max}(A_\rho) = \mathbb{T}$ and that A_ρ satisfies Condition (*). The sequence ρ is said to be of slow growth if $\rho_n = O(|n|^k)$ for some k>0. Finally we recall that a sequence $\{a_n\}_{n\in\mathbb{Z}}$ of numbers is dominated by a sequence $\{b_n\}_{n\in\mathbb{Z}}$

of positive numbers if there exists a constant c such that $|a_n| \leq \leq cb_n$, $n \in \mathbb{Z}$.

THEOREM [3, Theorem 1 a, b, c]. *Let* A *in* $L(E)$ *and suppose that there exist nonzero sequences* $\{x_n\}_{n\in\mathbb{Z}} \subset E$ *and* $\{y_n\}_{n\in\mathbb{Z}} \subset E^*$ *such that* $Ax_n = x_{n+1}$ *and* $A^*y_n = y_{n+1}$, $n \in \mathbb{Z}$ *and such that* $||x_n||_{n\in\mathbb{Z}}$ *and* $||y_n||_{n\in\mathbb{Z}}$ *are dominated by Beurling sequences. Then* A *has a nontrivial hyperinvariant subspace in each of the following two cases:*

1) $\sigma_A(x_o) \cup \sigma_{A^*}(y_o)$ *is not a singleton,*

2) *one of the Beurling sequences is of slow growth and* A *is not a multiple of the identity.*

REMARKS (1). We did not put explicitly in the hypothesis that A and A^* have the SVEP, thus it is a priori abusive to talk about $\sigma_A(x)$ and $\sigma_{A^*}(y)$. However an operator which does not satisfy the SVEP has eigenvalues and therefore there is no loss of generality with respect to the existence of hyperinvariant subspace for A in assuming that both A and A^* satisfy the SVEP.

(2) Atzmon's formulation was different but it is easily seen that $\sigma_A(x_o)$ and $\sigma_{A^*}(y_o)$ are the singular sets of the analytic functions G_1 and G_2 that he introduced.

PROOF. Let ρ_1 and ρ_2 denote the Beurling sequences which dominate respectively $||x_n||_{n\in\mathbb{Z}}$ and $||y_n||_{n\in\mathbb{Z}}$.

Let $B_i = A_{\rho_i}$, $i=1,2$ and let $T_1 \in L(B_1, E)$, $T_2 \in L(B_2, E^*)$ be defined by

$$T_1(f) = \sum_{n\in\mathbb{Z}} \hat{f}(n) x_n ,$$

$$T_2(g) = \sum_{n\in\mathbb{Z}} \hat{g}(n) y_n .$$

(That T_i, $i=1,2$, are bounded is an immediate consequence of the facts that ρ_1 dominates $||x_n||$ and that ρ_2 dominates $||y_n||$.)

The intertwinings $AT_1 = T_1U_1$ and $A^*T_2 = T_2U_2$ are obvious.

Since (Cf. Proposition 4.1) $\Sigma(T_1)\supset\sigma_A(T_1(1))=\sigma_A(x_o)$ and similarly $\Sigma(T_2)\supset\sigma_{A^*}(T_2(1))=\sigma_{A^*}(y_o)$ the set $\Sigma(T_1)\cup\Sigma(T_2)$ contains more than one point; by Theorem 3.1 A has a nontrivial hyperinvariant subspace. In the case where one of the Beurling sequences is of slow growth the corresponding Beurling algebra satisfies the singleton ideal structure condition of Theorem 3.2 and the second part of the conclusion follows from this same theorem.

REMARK. In general we are unable to deal with the case when $\sigma_A(x_o)$ and $\sigma_{A^*}(y_o)$ are both the same singleton by the above method while [3] contains partial results in that case.

6. INTERTWININGS AND THE S.BROWN TECHNIQUE

In this section we recall how the method was used to make reductions to obtain invariant subspaces for certain operators via the S.Brown technique [2]. Let A be an operator acting on a complex, separable, infinite dimensional Hilbert space H and let G be a bounded open set such that G^- is an M-spectral set for A, that is $||r(A)||\leq M \sup|r(\lambda)|$ for any rational function r with poles off G^-. This implies the existence of a norm-continuous representation Φ of $R(G^-)$ into $L(H)$ such that $\Phi(z)=A$. The application of the S.Brown method (see [5], Theorem 3.2) requires that 1) the extension of Φ to $H^\infty(G)$, the algebra of bounded analytic functions on G, and 2) the weak*-SOT sequential continuity of the extension (still denoted by Φ); that is whenever a bounded sequence $\{f_n\}_{n\in\mathbb{N}}\subset H^\infty(G)$ tends pointwise to 0 on G the corresponding sequence of operators $\{\Phi(f_n)\}$ tends to zero in the strong operator topology (i.e. $||\Phi(f_n)x||\to 0$, $x\in H$). It turns out that this latter property (restricted to sequences of $R(G^-)$) is sufficient to build a norm-continuous extension of Φ to $H^\infty(G)$ (provided any f in $H^\infty(G)$ is a pointwise limit of a sequence $\{f_n\}\subset R(G^-)$ bounded by $||f||$). Also it is enough to have either Φ or Φ^* (the adjoint representation of Φ; see below) satisfy the above continuity property to apply the S.Brown technique and obtain nontrivial invariant subspaces for A. The idea of the reduction is that the weak*-SOT sequential discontinuity of Φ implies the existence a nontrivial intertwining between M_z (acting on $R(\partial G)$) and A^*. Here for convenience to avoid "mixture" of Banach space duality and Hilbert space duality we will consider A as acting on a separable reflexive complex Banach space E. Thus in what follows Φ is a representation of a Banach algebra A (which will be $R(G^-)$ or $H^\infty(G)$) into $L(E)$. The adjoint representation Φ^* from A into $L(E^*)$ is defined by $\Phi^*(f)=(\Phi(f))^*$.

THEOREM 6.1. *Suppose that* Φ^* *is not weak*-SOT sequentially continuous and that* A *has no eigenvalues; then there exists a nonzero operator* $T:R(\partial G)\to E$ *intertwining* M_z *with* A.

PROOF. Since Φ^* is not w*-SOT sequentially continuous there exist a sequence f_n converging weak* to 0 in $H^\infty(G)$ and a vector y in E^* such that $||\Phi^*(f_n)y||$ does not tend to zero. By dropping to a subsequence we can assume that $\lim||\Phi^*(f_n)y||=\ell>0$. Choose unit vectors x_n in E such that $<x_n,\Phi^*(f_n)y>=||\Phi^*(f_n)y||$ that is $<\Phi(f_n)x_n,y>=||\Phi^*(f_n)y||$. Since E is reflexive any closed ball in E is weakly compact and metrizable and therefore, dropping again to a subsequence, we may assume that $\Phi(f_n)x_n$ is weakly convergent to say x. Clearly $<x,y>=\ell$ and therefore $x\neq 0$.

From now on the proof closely resembles the end of that of Theorem 2.1 of [2] (what has just been proved corresponds to Lemma 2.2 there); we just sketch it.

For a polynomial p we set $T(p)=p(A)x$. If (p/q) is a rational function with poles off ∂X then by Lemma 2.2 of [2] there exists a sequence of polynomials P_n such that $||P_n||_\infty\to 0$ $(p/q)(f_n-P_n)$ is a sequence in $R(G^-)$ tending weak* to 0. Next one shows easily that any weakly convergent subsequence of $v_n=\Phi(p/q)(f_n-P_n)x_n$ satisfies $q(A)(\lim v_n)=p(A)x$. Since we assume A to have no eigenvalues we conclude that v_n is weakly convergent and we set

$$T(p/q)=\lim v_n \quad .$$

One easily sees that $||T(p/q)||\leq||\Phi||(\limsup||f_n-P_n||)||(p/q)||_{\partial X}$ thus T can be extended to a bounded (linear) map $R(\partial G)\to E$. The intertwining relationship $TM_z=AT$ is verified first on p/q and then extended by continuity to all of $R(\partial G)$.

THEOREM 6.2. *Suppose that neither* Φ *nor* Φ^* *are weak**-SOT *sequentially continuous and that* $R(\partial G)=C(\partial G)$; *then* A *has a nontrivial hyperinvariant subspace.*

PROOF. We can assume without loss of generality that neither A nor A^* have eigenvalues. By the previous theorem since Φ^* is not weak*-SOT sequentially continuous there exists a nonzero bounded linear operator $T:R(\partial G)\to E$ intertwining M_z with A. Similarly (recall that E is reflexive) the weak*-SOT sequential discontinuity of Φ implies the existence of a nontrivial intertwining T' of M_z with A^*. Since any ideal in $C(X)$ whose hull is a singleton is a hyperplane the conclusion follows from Theorem 3.2.

REMARKS. The conclusion of Theorem 6.2 does not require the full strength of the hypothesis $R(\partial G)=C(\partial G)$. The following conditions would be sufficient:

a) $R(\partial G)$ is regular, and

b) any ideal I of $R(\partial G)$ whose hull $h(I)$ is a singleton is finite codimensional.

This suggests the corresponding natural questions: give conditions on G equivalent to conditions a) and b). As far as we know it may be the case that a) and b) are true for any bounded open set G.

REFERENCES

1. Apostol, C.: Teorie spectrală şi calcul funcţional, *St.Cer. Mat.* 20(1968), 635-668.

2. Apostol, C.; Chevreau, B.: On M-spectral sets and rationally invariant subspaces, *J.Operator Theory*, 7(1982), 247-266.

3. Atzmon, A.: On the existence of hyperinvariant subspaces, preprint.

4. Beauzamy, B.: Sous-espaces invariants de type fonctionnel, *Acta Math.* 144(1980), 65-82.

5. Chevreau, B.; Pearcy, C.; Shields, A.: Finitely connected domains G, representations of $H^\infty(G)$, and invariant subspaces, *J.Operator Theory*, 6(1981), 375-405.

6. Colojoară, I.; Foiaş, C.: *Theory of generalized spectral operators*, New York, London, Gordon and Breach, 1968.

7. Gellar, R.; Herrero, D.: Hyperinvariant subspaces of bilateral

weighted shifts, *Indiana Univ.Math.J.* 23(1974), 771-790.

8. Katznelson, Y.: *An introduction to harmonic analysis*, Dover Publications, New York.

9. Lumer, G.; Rosenblum, M.: Linear operators equations, *Proc. Amer.Math.Soc.* 10(1959), 32-41.

10. Sz.Nagy, B.; Foiaş, C.: *Harmonic analysis of operators on Hilbert space*, Akademiai Kiado Budapest, 1970.

11. Radjabalipour, M.: Decomposable operators, *Bull.Iranian Math. Soc.* 9(1978), 1-49.

12. Wermer, J.: The existence of invariant subspaces, *Duke Math. J.* 19(1952), 615-622.

B.Chevreau
Department of Mathematics,
INCREST,
Bd.Păcii 220,
79622 Bucharest,
Romania.

Present address:

UER de Mathématique et d'Informatique,
Université de Bordeaux I,
351, Cours de la Libération,
33405 Talence,
France.

ON MODULI FOR INVARIANT SUBSPACES

M.J.Cowen and R.G.Douglas

One of the most influential papers in operator theory was that of Beurling [4] in which he proposed to solve the problem of spectral synthesis for the backward shift operator on the Hilbert space ℓ^2. His characterization of the adjoint operator's invariant subspaces in terms of inner functions and his introduction of the notion of inner-outer factorization into the problem had a profound effect.

On the Hilbert space

$$\ell^2=\{(\alpha_0,\alpha_1,\alpha_2,\ldots):\alpha_n\in\mathbb{C},\Sigma|\alpha_n|^2<\infty\}$$

the shift operator S is defined by

$$S(\alpha_0,\alpha_1,\alpha_2,\ldots)=(0,\alpha_0,\alpha_1,\alpha_2,\ldots).$$

This mapping of an orthonormal basis onto a proper subset of itself captures in the simplest possible way the essence of the new phenomena that can occur on an infinite dimensional space. Although the definition of this operator must have occurred to many early researchers, the basic results about it were established by von Neumann [10] in connection with his characterization of maximally symmetric operators.

Recall that an isometry is an operator V defined on some Hilbert space H satisfying $||Vx||=||x||$ for all vectors in H. It is unitary if it is onto, in which case the spectral theorem applies. The basic result concerning isometries (and the shift operator) is the von Neumann-Wold decomposition [10],[11].

THEOREM. *If* V *is an isometry on the complex Hilbert space* H, *then there exists a unique decomposition* $H=H_u\oplus H_s$ *into reducing subspaces for* V *such that*

(1) $V|H_u$ *is unitary, and*

(2) $V|H_s$ *is the direct sum of* n *copies of* S, *where* $n=\dim(H\ominus VH)$.

Now, as is known (cf.[7]), the Beurling description of the invariant subspaces of S can be obtained from this result in a rather straightforward manner. Let M be a closed subspace of ℓ^2 such that $SM \subset M$. The operator V defined on M by $V=S|M$ is an isometry, has no unitary part since

$$\bigcap_{n\geq 0} V^n M \subset \bigcap_{n\geq 0} S^n \ell^2 = \{0\},$$

and a multiplicity argument shows that V is unitarily equivalent to S unless $M=\{0\}$. If $U:\ell^2 \to M$ is a unitary operator which implements this equivalence, then $VU=US$ and if the operator $\tilde{U}$ is defined on ℓ^2 to be U followed by the inclusion of M in ℓ^2, then $\tilde{U}S=S\tilde{U}$. Thus $\tilde{U}$ is an isometry commuting with S such that $M=\tilde{U}\ell^2$. If we transfer this consideration to the Hardy space H^2 using Fourier series, then we obtain the usual form of Beurling's result; namely, the invariant subspaces of multiplication by z on H^2 are of the form ϕH^2 for some inner function ϕ. This results from the fact that an operator which commutes with multiplication by z must be multiplication by a bounded holomorphic function on the open unit disk and must have boundary values of modulus one a.e. if an isometry.

What is the point of our presenting the foregoing proof? Certainly there are shorter proofs of Beurling's result (cf.[8]) and indeed, there are a couple of facts used in the above, that while elementary, are not completely trivial. Our point in presenting this is the belief that the basic result is just the fact that the collection of operators obtained by restricting S to its nonzero invariant subspaces are all unitarily equivalent to S.As the proof indicates, this fact gives rise to the isometry representing an invariant subspace and the particular form Beurling's theorem takes results from the simple characterization of the operators which commute with S. And while the Beurling characterization yields a set theoretic description of an invariant subspace M as the functions $\{\phi f: f\in H^2\}$, many applications use only the fact that S on ℓ^2 is a model for $S|M$.[1)]

[1)] This is not always the case since sometimes the lattices structure is needed which Beurling's description provides.

In only a few instances is it true[2)] that all the restriction operators for a given operator are pairwise unitarily equivalent. Thus the analogue of Buerling's theorem can hold for few operators and hence "inner functions" will not suffice in contexts where the commutant is given by multipliers. This explains the failure of various attempts at generalizing Beurling's theorem. However, the problem of seeking models for the restriction operators or a space of moduli, to borrow a term from algebraic geometry, makes sense and indeed was solved in an interesting case by Abrahamse and the second author [1].

Let Ω be a finitely connected domain in $\mathbb{C}$, μ harmonic measure on $\partial\Omega$ for some interior point ω_o in Ω, and $H^2(\Omega)$ the closure in $L^2(\mu)$ of the rational functions $\mathrm{Rat}(\Omega)$ with poles off the closure of Ω. One considers not only the operator defined to be multiplication by z on $H^2(\Omega)$ but rather the whole algebra of operators obtained by representing $\mathrm{Rat}(\Omega)$ on $H^2(\Omega)$ as multiplication operators:

$$\phi \to T_\phi \qquad \text{on } H^2(\Omega).$$

One is interested in the algebra representations of $\mathrm{Rat}(\Omega)$ obtained by restricting

$$\phi \to T_\phi | M \qquad \text{for } \phi \text{ in } \mathrm{Rat}(\Omega)$$

to a subspace M which is invariant for the whole algebra of operators. In case Ω is the unit disk then this is solved by the von Neumann-Wold decomposition. For the general case Abrahamse and the second author showed how to describe models for the restrictions in terms of a certain class of vector bundles over Ω.

If E is a flat unitary holomorphic line bundle over Ω, then we can form a Hardy space $H^2_E(\Omega)$ from its square-integrable holomorphic cross-sections on which $\mathrm{Rat}(\Omega)$ acts by pointwise multiplication.

THEOREM. *Let* M *be a subspace of* $H^2(\Omega)$ *which is invariant for the algebra of operators* $\{T_\phi : \phi \in \mathrm{Rat}(\Omega)\}$. *Then there exists a flat unitary holomorphic line bundle* E *over* Ω *and an isometric operator*

2) Indeed, assuming that all operators have invariant subspaces, it seems likely that there are no others not essentially related to the shift.

$$U: H^2_E(\Omega) \to H^2(\Omega)$$

such that (1) $UH^2_E(\Omega) = M$ *and* (2) $UT_\phi = T_\phi U$ *for* ϕ *in* Rat(Ω). *Moreover, all such bundles occur and the bundle is uniquely determined up to equivalence.*

Thus in the language we introduced earlier a moduli space for the restriction representations of Rat(Ω) is provided by the collection of these line bundles over Ω. As one can show the moduli space for such bundles is given by the character group of the fundamental group for Ω. In particular, the moduli space for Rat(Ω) acting on $H^2(\Omega)$ is the n-torus, where n is the connectivity of Ω. Thus not only does one obtain a concrete description for the moduli space, but in this case there is a "natural" topology relative to which the space is a compact manifold. Analogous results were obtained for the case of higher multiplicity, that is, for Rat(Ω) acting on the direct sum of k copies of $H^2(\Omega)$ for $1 \le k \le \infty$; the moduli space is

$$\underbrace{U_k \times U_k \times \cdots \times U_k}_{\text{n copies}} / U_k ,$$

where U_k is the unitary group on the Hilbert space of dimension k and the action of U_k on the n-tuples is by conjugation.

If instead of unitary equivalence, one considers similarity, then the moduli space consists of a single point for each possible multiplicity. Thus two restriction algebras Rat$(\Omega) | H^2_{E_0}(\Omega)$ and Rat$(\Omega) | H^2_{E_1}(\Omega)$ are similar if and only if the ranks of E_0 and E_1 are equal. This results from a theorem of Grauert and Bungart.

If one considers more or less arbitrary operators or algebras of operators, then it is unreasonable to expect a good moduli space to exist for the restriction operators, or indeed that one should be able to provide a complete set of models. However, for natural examples of operator algebras, good moduli spaces should exist both for unitary equivalence and similarity. In some cases it may be necessary to consider moduli spaces in which different points may yield equivalent models.

What are some "natural examples" that one might consider? The most likely to produce good results, we believe, would be representations of algebras of holomorphic functions as multiplication operators—in other words, the same kind of spaces we have been discussing. Obvious choices would be the algebra $\mathrm{Rat}(\Omega)$ acting on other Hilbert spaces and a good place to start would be with multiplication by z on the Bergman space. Although some results are known [9], the problem has not been considered from this point of view. We'll have more to say about this later. Other important examples worth considering are the algebra of holomorphic functions on the poly disk or the unit ball in $\mathbb{C}^n$ acting as multiplication on either the Hardy or Bergman spaces. These problems are undoubtedly difficult but we believe the time is right for considering them.

If one is considering an algebra of holomorphic functions on a domain in $\mathbb{C}$ or $\mathbb{C}^n$ which is not simply connected, then flat unitary holomorphic vector bundles should play a role and may reduce the problem to the simply connected case. Anyhow we shall concentrate on the latter case from here on.

How might one obtain models? One possibility is to consider ideals in the algebra. For example let I be an ideal in the algebra $\mathbb{C}[z_1,z_2]$ of polynomials in two complex variables. If $[I]$ denotes its closure in the Hardy space $H^2(\mathbb{D}^2)$, then $[I]$ is invariant for $\mathrm{Rat}(\mathbb{D}^2)$ and hence $\mathrm{Rat}(\mathbb{D}^2)|[I]$ is a restriction representation of $\mathrm{Rat}(\mathbb{D}^2)$. Thus the ideals in $\mathbb{C}[z_1,z_2]$ provide models for the restriction of $\mathrm{Rat}(\mathbb{D}^2)$ to some invariant subspaces—does it for all? This is not known but it doesn't seem unreasonable. If one considers the variety $z(I)$ formed by the common zeros of the polynomials in I, then the complex dimension of $z(I)$ is either zero or one. In the zero-dimensional case, the variety consists of a finite discrete set of points. A result of Ahern and Clark [2] can be rephrased to yield.

THEOREM. *If M is a subspace of finite codimension in* $H^2(\mathbb{D}^2)$ *which is invariant for multiplication by* $\mathrm{Rat}(\mathbb{D}^2)$, *then there exists an ideal I in* $\mathbb{C}[z_1,z_2]$ *such that* $M=[I]$.

It is not known that different ideals give rise to different restriction algebras. Further, the zero-dimensional ideals model more than the invariant subspaces of finite codimension since one can multiply the invariant subspace by an inner function without changing the restriction algebra. An intrinsic characterization of the invariant subspaces modeled in this case would be most interesting. Some notion of module dimension might be relevant. An investigation of the restrictions to ideals having a one-dimensional variety would also be extremely interesting since entirely new phenomena will be encountered here. (In the classical situation of planar algebras, all ideals have zero-dimensional variety.)

As we said the moduli space in the zero-dimensional case may be only a resolution and hence it is important to determine for which pairs of ideals the restrictions are equivalent. What is known about when ideals give rise to inequivalent models? Consider the following ideal $I_{p,q}$ having zero variety consisting of just the origin in the bidisk: the Taylor coefficients α_{ij} in the Taylor series expansion $\Sigma\alpha_{ij}z_1^i, z_2^j$ about the origin for a function f in $I_{p,q}$ are required to vanish for $0 \le i \le p$, $0 \le j \le q$. In [3] Berger, Coburn, and Lebow showed that all the restriction algebras are inequivalent. They obtained this result for finite intersections of such ideals. In [6] we gave an alternate proof of this result by showing that the adjoints of the restriction of the two shift operators to the corresponding invariant subspaces form a pair of operators to which the several variables analogue of our complex geometric techniques [5] applied. In particular, the curvature of the appropriate Hermitian holomorphic line bundle was shown to have encoded in it the p's and q's describing the ideal. We expect that these techniques can be extended to cover all ideals having a single point as its zero variety and a student Agrawal has made some progress on this. Actually this method may be sufficient to resolve completely the case of ideals with zero-dimensional zero varieties. His results suggest that in the zero-dimensional case distinct ideals yield inequivalent models.

The prospect of obtaining models from ideals is, in general, very exciting not because that would solve the problem. Indeed, the ideal structure of most algebras is quite complicated. However, a bridge between the two problems would be useful for both subjects. The use of the algebra $\mathbb{C}[z_1,z_2]$, as we indicated, is probably only a crude first approximation since one might use it equally well for both the bidisk and the unit sphere. However, we would expect the results for these two cases to be quite different.

It may be possible to use techniques from complex geometry in other ways in studying invariant subspace restrictions. Recall that for Ω a bounded open connected subset of $\mathbb{C}$ we define the class of operators $\mathcal{B}_1(\Omega)$ to be those T on H which satisfy 1) $\Omega\subset\sigma(T)$, 2) $(T-\omega)H=H$ for ω in Ω, 3) $\bigvee_{\omega\in\Omega}\ker(T-\omega)=H$, and 4) $\dim\ker(T-\omega)=1$. If M is an invariant subspace for T^*, then it seems likely that the compression $P_M T|M$ also lies in $\mathcal{B}_1(\Omega)$. That this is true has been claimed by Morrel but we can prove it only under the additional assumption that $\dim(H\ominus M)<\infty$.

Let ω_o be a point in Ω and let

$$T=\begin{bmatrix} A & D \\ B & C \end{bmatrix}$$

be the decomposition relative to $H=M\oplus M^\perp$. Hence, $A=P_M T|M$. If $T-\omega_o$ is onto, then it is immediate that $A-\omega_o$ is onto. If the vector $x\oplus y$ lies in $\ker(T-\omega_o)$, then x lies in $\ker(A-\omega_o)$. However, it is possible that x=0. Thus let $x(\omega_o)$ be a holomorphic cross-section for $\ker(T-\omega)$. Then $Ax(\omega)=\omega x(\omega)$. Since $x(\omega)\equiv 0$ contradicts the assumption that the kernels span H, we have $x(\omega)\not\equiv 0$. Therefore, there exists a holomorphic vector-valued function $\tilde{x}(\omega)$ such that $x(\omega)=(\omega-\omega_o)^k\tilde{x}(\omega_o)$ and $\tilde{x}(\omega_o)\neq 0$. Since

$$A((\omega-\omega_o)^k\tilde{x}(\omega))=\omega(\omega-\omega_o)^k\tilde{x}(\omega)$$

it follows that $A\tilde{x}(\omega)=\omega\tilde{x}(\omega)$ for $\omega\neq\omega_o$ and hence for all ω. Thus $\ker(A-\omega_o)$ is at least one dimensional. Therefore, we see that $P_M T|M$ satisfies 1), 2), and 3) and $\dim\ker(P_M|M-\omega I_M)\geq 1$. Our difficulty lies in showing that the latter subspaces have dimension equal to one.

If $N=\ker(A-\omega)$, then $(T-\omega)|N\oplus M^\perp$ has range equal to $M^\perp$ and has kernel equal to $\ker(T-\omega)$. Thus, if $M^\perp$ is finite dimensional, then $\dim N$ equals $\dim\ker(T-\omega)$ and hence $P_M T|M$ lies in $\mathcal{B}_1(\Omega)$.

The problem now is to relate the curvature of the restriction operators $P_M T|M$ to the curvature K_T of T. There is some limited evidence to suggest for T^* subnormal (which all the operators we have been discussing are), that there may be a close relation and, in particular, one may be able to limit the possible curvatures for the restrictions. One possibility which may need additional hypotheses is that

$$0 > K_{P_M T|M} \geq K_T$$

(perhaps only asymptotically as one approaches the boundary of Ω). We point out that it would be worthwhile to examine the Bergman shift from this point of view.

We will speculate no further. These are very interesting problems likely to make contact with the rest of mathematics and ripe for investigation.

REFERENCES

1. Abrahamse, M.B.; Douglas, R.G.: A class of subnormal operators related to multiply-connected domains, *Advances in Math.* 19(1976), 106-148.

2. Ahern, P.; Clark, D.: Invariant subspaces and analytic continuation in several variables, *J.Math.Mech.* 19(1969/70), 963-969.

3. Berger, C.A.; Coburn, L.A.; Lebow, A.: Representation and index theory for C*-algebras generated by commuting isometries, *J.Functional Analysis* 27(1978), 51-99.

4. Beurling, A.: On two problems concerning linear transformations in Hilbert space, *Acta Math.* 81(1949), 239-255.

5. Cowen, M.J.; Douglas, R.G.: Complex geometry and operator theory, *Acta Math.* 141(1978), 187-261.

6. Cowen, M.J.; Douglas, R.G.: On operators possessing an open set of eigenvalues, *Proceedings Symposium for Fejer-Riesz*, to appear.

7. Fillmore, P.: *Notes on operator theory*, van Nostrand, New York, 1970.

8. Helson, H.: *Lectures on invariant subspaces*, Academic Press, New York, 1964.

9. Horowitz, C.: Zeros of functions in the Bergman spaces, *Duke Math.J.* 41(1974), 693-710.

10. von Neumann, J.: Allgemeine Eigenwerttheorie Hermitescher Funktional Operatoren, *Math.Ann.* 102(1929), 49-131.

11. Wold, H.: *A study in the analysis of stationary time series*, Stockholm, 1938.

M.J.Cowen
Department of Mathematics,
State University of New York
at Buffalo,
Buffalo, NY 14214,
U.S.A.

R.G.Douglas
Department of Mathematics,
State University of New
York at Stony Brook,
Stony Brook, NY 11794,
U.S.A.

AN EXTENSION OF CHACON-ORNSTEIN ERGODIC THEOREM

Radu-Nicolae Gologan

Let $(X,\mathcal{X},\mu)$ be a σ-finite measure space. If P is an $L^1(X,\mathcal{X},\mu)$ positive linear contraction (i.e. $Pf\geq 0$ for $f\in L^1$, $f\geq 0$ and $\int|Pf|d\mu\leq\leq\int|f|d\mu$ for $f\in L^1$) the celebrated theorem of Chacon and Ornstein ([3]) asserts that for $f,g\in L^1$, $g\geq 0$:

$$\lim_{n\to\infty}\frac{f+Pf+\ldots+P^n f}{g+Pg+\ldots+P^n g}$$

exists and is finite μ-a.e. on the set $\{x|\sum_{i=0}^{\infty}(P^i g)(x)>0\}$. In [2] Chacon gives an extension of this result for nonpositive contractions: if T is an arbitrary L^1-contraction, $f\in L^1$ and $\{p_n\}_{n\in\mathbb{N}}$ is a sequence of positive measurable functions such that for every $n\in\mathbb{N}$, $g\in L^1_+$, $g\leq p_n$ implies $|Tg|\leq p_{n+1}$, then

$$\lim_{n\to\infty}\frac{f+Tf+\ldots+T^n f}{p_o+p_1+\ldots+p_n}$$

exists and is finite μ-a.e. on the set $\{x|\sum_{i=0}^{\infty}p_i(x)>0\}$.

A different aproach was given by Cuculescu and Foiaş in [4]. They prove that under the same hypothesis:

$$\lim_{n\to\infty}\frac{|f|+|Tf|+\ldots+|T^n f|}{p_o+p_1+\ldots+p_n}$$

exists and is finite μ-a.e. on the same set.

Using the technique of Cuculescu and Foiaş based on a construction of Brunnel ([1]) we shall give conditions under which the limit (whose existence follows from the above theorem):

$$\lim_{n\to\infty}\frac{|f|+|Tf|+\ldots+|T^n f|}{|g|+|Tg|+\ldots+|T^n g|}$$

for $f,g \in L^1$, is finite μ-a.e.. In the particular case when T is conservative we shall describe the limit.

In stating our result we shall recall some results from [4] and [5].

Let P be a $L^1(X,\mathcal{X},\mu)$ positive contraction and P^* its adjoint acting in $L^\infty(X,\mathcal{X},\mu)$. For $A \in \mathcal{X}$ let us denote by χ_A the caracteristic function of A.

PROPOSITION 1. ([4]). *With the above notations the function*

$$e_A = \inf\{g \in L^\infty \mid g \geq \chi_A \ , \ P^*g \leq g\}$$

is in L^∞ *and satisfies* $P^*e_A \leq e_A$.

The function e_A is called the *equilibrium potential* of A.

For $f \in L^1$, f a real function, we shall denote by f^+, respectively by f^-, the positive, resp. the negative part of f.

The following is a version of the maximal ergodic lemma of Brunnel, Cuculescu and Foiaş ([1],[4]):

PROPOSITION 2. *Let* P *be a positive linear contraction in* L^1, $\{g_n\}_{n \in \mathbb{N}}$ *a sequence of real* L^1 *functions and* A *a measurable set such that*:

$$A \subset \{x \mid \sup_{n \geq k} \sum_{i=k}^{n} g_i(x) > 0, \textit{ for every } k \geq 0\}.$$

Then:

$$\int [g_o e_A + \sum_{i=1}^{\infty} (g_i - Pg_{i-1})^+ e_A] d\mu \geq \int [\sum_{i=1}^{\infty} (g_i - Tg_{i-1})^- e_A] d\mu \ .$$

The following result of Krengel is now well-known:

PROPOSITION 3. *Let* T *be a linear contraction in* L^1. *There exists a positive linear contraction* $|T|$ *such that*:

$$|Tf| \leq |T|\,|f| \qquad \textit{for every } f \in L^1.$$

Let us recall that a linear contraction T is called conservative if for every $f \in L^1_+$ we have:

$$\sum_{n=0}^{\infty} |T|^n f = 0 \text{ or } \infty \ \ \mu\text{-a.e.}$$

A set $A\in X$ is called T-invariant if $\mathrm{supp}|Tf|\subset A$ whenever $\mathrm{supp}\, f\subset A$ for $f\in L^1_+$. It is well-known that if T is conservative the class I_T of T-invariant sets is a σ-algebra.

We shall state our Chacon-Ornstein type theorem in a general form. To this end we need some simple results which we formulate as a lemma.

LEMMA. *Let* T *be a* L^1*-linear contraction and* $\{p_n\}_{n\in\mathbb{N}}\subset L^1_+$ *a sequence of positive functions such that* $|Tp_n|\geq p_{n+1}$ *for every* $n\in\mathbb{N}$. *Then, denoting by* e_A *the equillibrium potential of* $A\in X$ *with respect to* $|T|$, *the limit:*

$$p(A)=\lim_{n\to\infty}\int e_A p_n d\mu$$

exists and it is finite.

Moreover, if T *is conservative* p *defines a measure on* I_T *which is absolutely continuous with respect to* μ.

PROOF. Using Propositions 1 and 3 we have:

$$\int e_A p_n d\mu \leq \int e_A |Tp_{n-1}| d\mu \leq \int e_A |T| p_{n-1} d\mu =$$

$$= \int |T|^* e_A p_{n-1} d\mu \leq \int e_A p_{n-1} d\mu$$

for every $n\in\mathbb{N}^*$. So the sequence defining p(A) is positive and decreasing.

The fact that in the case T conservative p defines an absolutely continuous measure on I_T is a simple consequence of the equality $e_A=\chi_A$ valid in this case for $A\in I_T$ (see [5]).

We need also the following:

DEFINITION. If $P=\{p_n\}_{n\in\mathbb{N}}$ is a sequence in L^1_+ such that $|Tp_n|\geq p_{n+1}$ for every $n\in\mathbb{N}$, a set $A\in X$, $\mu(A)>0$, is called *P-acceptable* if:

$$\lim_{n\to\infty}\int e_{A'} p_n d\mu=0 \qquad \text{for } A'\subset A$$

implies $\mu(A')=0$.

It is easy to see that in the conservative case this means that μ and p are equivalent on A.

We can now formulate our result.

THEOREM. *Let* T *be a linear contraction in* $L^1(X,\mathcal{X},\mu)$ *and* $P=\{p_n\}_{n\in\mathbb{N}}$, $Q=\{q_n\}_{n\in\mathbb{N}}$ *two sequences in* L^1_+ *such that* $|Tp_n|\geq p_{n+1}$ *and* $|Tq_n|\geq q_{n+1}$ *for every* $n\in\mathbb{N}$, $(q_o\neq 0)$. *Then*:

$$\lim_{n\to\infty}\frac{p_o+p_1+\ldots+p_n}{q_o+q_1+\ldots+q_n} \quad \text{exists } \mu\text{-a.e.},$$

and is finite on every Q*-acceptable set.*

Moreover, in the case T *is conservative, on every* Q*-acceptable set this limit is* $E(P)/E(Q)$ a.e., *where* $E(P)$, *resp.* $E(Q)$, *are the Radon-Nikodym derivatives of the measures described above.*

PROOF. The existence of the limit is a direct consequence of the ergodic theorem of Cuculescu and Foiaş.

Let us prove that the limit is finite on every Q-acceptable set $B\in\mathcal{X}$.

Suposing the contrary it follows that for every $\alpha>0$ and $k\in\mathbb{N}$:

$$\sup_{n\geq k}\sum_{i=k}^{n}(p_i-\alpha q_i)>0 \text{ on } B.$$

Denoting by $h_i=p_i-\alpha q_i$ and using Proposition 2 we have:

$$(1)\qquad \int[e_B h_o+\sum_{i=1}^{\infty}(h_i-|T|h_{i-1})^+ e_B]d\mu\geq\int[\sum_{i=1}^{\infty}(h_i-|T|h_{i-1})^- e_B]d\mu\geq 0.$$

From Proposition 3 and the conditions on P and Q we have:

$$(h_i-|T|h_{i-1})^+=[p_i-|T|p_{i-1}+\alpha(|T|q_{i-1}-q_i)]^+\leq$$

$$\leq\alpha(|T|q_{i-1}-q_i).$$

We have then from (1):

$$0\leq\int[e_B(p_o-\alpha q_o)+\alpha\sum_{i=1}^{\infty}e_B(|T|q_{i-1}-q_i)]d\mu=$$

$$=\lim_{n\to\infty}\int[e_B(p_o-\alpha q_o)+\alpha\sum_{i=1}^{n}(|T|^*e_B q_{i-1}-q_i e_B)]d\mu\leq$$

$$\leq\lim_{n\to\infty}\int[e_B(p_o-\alpha q_o)+\alpha\sum_{i=1}^{n}(e_B q_{i-1}-e_B q_i)]d\mu=$$

$$= \int e_B p_o d\mu - \alpha \lim_{n\to\infty} \int e_B q_n d\mu \quad ,$$

where we have used the inequality $|T|^* e_B \le e_B$. As α was arbitrary and $B \subset A$ it follows $\mu(B)=0$.

Let us prove the second half of the theorem. A being $\mathcal{Q}$-acceptable and T conservative, denoting by:

$$r_n = \frac{p_o + p_1 + \ldots + p_n}{q_o + q_1 + \ldots + q_n}$$

it will suffice to prove that for $0 \le a < b$ on the set $B = \{x \mid a < \lim r_n < b\} \cap A$ we have $\alpha \le E(\mathcal{P})/E(\mathcal{Q}) \le b$ μ-a.e. The function $E(\mathcal{P})/E(\mathcal{Q})$ being $\mathcal{I}_T$-measurable we shall prove the last inequality on the T-invariant closure $I(B)$ of B.

Let $M \in \mathcal{I}_T$, $M \subset I(B)$. From [5] we know that $e_{M\cap B} = e_B$. Using Proposition 2, in the same way as in the first part of the proof, we obtain:

$$\int [e_M p_o + \sum_{i=1}^{\infty} (p_i - |T| p_{i-1}) e_M)] d\mu \ge \int [e_M q_o + \sum_{i=1}^{\infty} (q_i - |T| q_{i-1}) e_M] d\mu$$

and

$$b \int [e_M q_o + \sum_{i=1}^{\infty} (q_i - |T| q_{i-1}) e_M] d\mu \ge \int [e_M p_o + \sum_{i=1}^{\infty} (p_i - |T| p_{i-1}) e_M] d\mu \quad .$$

Using the preceeding lemma and the fact that $M \in \mathcal{I}_T$ implies $|T|^* e_M = e_M$ (see [5]) we conclude:

$$\int_M (E(\mathcal{P}) - aE(\mathcal{Q})) d\mu \ge 0$$

and

$$\int_M (bE(\mathcal{Q}) - E(\mathcal{P})) d\mu \ge 0 \quad .$$

M being an arbitrary $\mathcal{I}_T$-set in $I(B)$ these inequalities imply the last conclusion of the theorem.

The following consequence can be viewed as a Chacon-Ornstein theorem for nonpositive contractions.

COROLLARY. *Let* T *be a linear contraction on* $L^1(X, \mathcal{X}, \mu)$ *and* $f, g \in L^1$, $g \ne 0$. *The limit:*

$$\lim_{n\to\infty} \frac{|f|+|Tf|+\ldots+|T^n f|}{|g|+|Tg|+\ldots+|T^n g|}$$

exists and is finite μ-a.e. *on every* $\{|T^n g|\}$*-acceptable set. Moreover, if* T *is conservative, on every such set the limit coincides with the quotient of the Radon-Nikodym derivatives of the measures defined on* I_T *by:*

$$f(A)=\lim_{n\to\infty} \int_A |T^n f|\, d\mu \ ,$$

$$g(A)=\lim_{n\to\infty} \int_A |T^n g|\, d\mu \ .$$

REFERENCES

1. Brunnel, A.: Sur une lemme ergodique voisine du lemme de Hopf et sur une de ses applications, *C.R.Acad.Sci.Paris* 256(1963), 5481-5484.

2. Chacon, R.V.: Convergence of operator averages, in "*Proc.internat.sympos.in ergodic theory*", Academic Press, New York, 1963.

3. Chacon, R.V.; Ornstein, D.S.: A general ergodic theorem, *Illinois Math.J.* 4(1960), 153-160.

4. Cuculescu, I.; Foiaş, C.: An individual ergodic theorem for positive operators, *Rev.Roumaine Math.Pures Appl.* 11(1966), 581-594.

5. Meyer, P.A.: Théorie érgodique et poténtiel. Part.I-II, *Ann. Inst.Fourier (Grenoble)* 15(1965), 581-594.

R.-N.Gologan
Department of Mathematics,
INCREST,
Bdul Păcii 220, 79622 Bucharest,
Romania.

COMMUTING SUBNORMAL OPERATORS QUASISIMILAR TO MULTIPLICATION BY COORDINATE FUNCTIONS ON ODD SPHERES

J. Janas

Let $S=(S_1,\dots,S_n)$ be a jointly subnormal n-tuple of pairwise commuting operators on a separable Hilbert space H. Denote by $N=(N_1,\dots,N_n)$ its commuting normal extension on a Hilbert space $K\supset H$, i.e.

$$S_i f=N_i f\ ,\qquad f\in H\ ,\ i=1,\dots,n\ .$$

Assume that there exists $f_o\in H$ which is cyclic for S. It means that the smallest subspace of H containning f_o and invariant under $S_1,\dots,S_n$ is all H. It turns out that such S has a special form. Namely,denote for a finite Borel measure μ in $\mathbb{C}^n$ by $H^2(\mu)$ the $L^2(\mu)$ closure of polynomials in $z=(z_1,\dots,z_n)$. Then by the result of [3] or [6] there exists a Borel measure ν with compact support in $\mathbb{C}^n$ and there exists a unitary operator $V:K\to L^2(\nu)$ such that $Vf_o=1$, $VH=H^2(\nu)$ and $VS_i=\mathcal{U}_{z_i}V$, where $\mathcal{U}_{z_i}f=z_if$, $f\in H^2(\nu)$, $i=1,\dots,n$.

Denote by $\mathbb{B}=\{z\in\mathbb{C}^n,|z|<1\}$-the open unit ball in $\mathbb{C}^n$. In the case $\nu=m$, where m is the surface Lebesgue measure on the topological boundary $\partial\mathbb{B}=\mathbb{S}$ of $\mathbb{B}$, $H^2(\nu)$ reduces to the classical Hardy space H^2 in the unit ball $\mathbb{B}$. Let T_{z_i} be the operator of multiplication by z_i on H^2. In what follows we shall investigate when an n-tuple $\mathcal{U}_\nu=(\mathcal{U}_{z_1},\dots,\mathcal{U}_{z_n})$ is quasisimilar to $T=(T_{z_1},\dots,T_{z_n})$. It means we shall look for operators $X:H^2(\nu)\to H^2$ and $Y:H^2\to H^2(\nu)$ with dense ranges and zero kernels such that

$$T_{z_i}X=X\mathcal{U}_{z_i}\ ,\qquad YT_{z_i}=\mathcal{U}_{z_i}Y\ ,\qquad i=1,\dots,n\ .$$

We shall write then $T \sim\sim \mathcal{U}_\nu$. We shall give a necessary and sufficient condition, in terms of the measure ν, which guarantees that $T \sim\sim \mathcal{U}_\nu$. In the case n=1, the above question has been answered completely by Clary [1]. The same problem also has been solved for the case of multiplication by coordinate functions on the Hardy space $H^2(\mathbb{D}^n)$ over the polydisc $\mathbb{D}^n \subset \mathbb{C}^n$, by Hastings [3]. Although our condition is not so satisfactory as that given by Clary, we shall give some examples which should clarify its meanning.

We say that $\phi \in H^2(\mu)$ is cyclic, if $P \cdot \phi$ is dense in $H^2(\mu)$, where P denotes the set of all polynomials. In other words ϕ is a joint cyclic vector for $\mathcal{U}_\mu$. We start with the following

PROPOSITION 1. *Let* $\mu \geq 0$ *be a finite Borel measure in* $\mathbb{C}^n$ *and* $\operatorname{supp}\mu \subseteq \overline{\mathbb{B}}$ *(the closure). If* $Y: H^2 \to H^2(\mu)$ *has a dense range and satisfies*

$$YT = \mathcal{U}_\mu Y \ ,$$

then

(i) $\mu | \mathbb{S} \ll m$;

(ii) *there is a cyclic vector* $\phi \in H^2(\mu)$ *such that*

$$\int |p\phi|^2 d\mu \leq \int_{\mathbb{S}} |p|^2 dm, \qquad \textit{for any polynomial } p.$$

PROOF. We can assume that $\|Y\|=1$. Let $\phi = Y1$. Define a measure ν on $\mathbb{S}$ by

$$\nu(E) = \int_E |\phi|^2 dm \ , \qquad E \subset \mathbb{S} \ .$$

Then for any polynomial p we have

$$\int_{\mathbb{S}} |p|^2 d\nu = \int_{\mathbb{S}} |p|^2 |\phi|^2 d\mu \leq \int_{\overline{\mathbb{B}}} |Yp|^2 d\mu \leq \int_{\mathbb{S}} |p|^2 dm \ .$$

Hence applying Lemma of [4]

$$\int_{\mathbb{S}} h d\nu \leq \int_{\mathbb{S}} h dm \ ,$$

for any positive and continuous function h on $\mathbb{S}$. Thus $\nu \ll m$. Now by our assumption $Yp=p\phi$, for any polynomial p. Since the range $R(Y)$ is dense in $H^2(\mu)$, ϕ is a cyclic vector and $\mu\{z\in\mathbb{C}^n,\ \phi(z)=0\}=$ $=0$. Therefore $\nu\sim\mu|S$ and the proof is complete.

We also have

PROPOSITION 2. *Let μ be a finite, positive Borel measure in $\mathbb{C}^n$ with* $\operatorname{supp}\mu\subseteq\bar{B}$. *Assume that there is $X:H^2(\mu)\to H^2$ with dense range such that $X\mathcal{U}_\mu=TX$. Then there exists a cyclic function $\psi\in H^2$ for which $\frac{d\mu_o}{dm}\geq|\psi|^2$ a.e. [m], where $\mu_o=\mu|S$.*

PROOF. As before we can assume that $||X||=1$. Let $\psi=X1$. Then $Xp=\psi p$, $p\in P$, so ψ is a cyclic vector. Now for any positive integer $s_i\geq 0$, $i=1,\dots,n$ and $f\in H^2(\mu)$

$$T_{z_1^{s_1}}\dots T_{z_n^{s_n}}Xf=X\mathcal{U}_{z_1^{s_1}}\dots\mathcal{U}_{z_n^{s_n}}f\ .$$

Hence for $g\in H^2$, $f\in H^2(\mu)$ we have

$$(T^*_{z_1^{s_1}}\dots T^*_{z_n^{s_n}}T_{z_1^{s_1}}\dots T_{z_n^{s_n}}Xf,g)=(T^*_{z_1^{s_1}}\dots T^*_{z_n^{s_n}}X\mathcal{U}_{z_1^{s_1}}\dots\mathcal{U}_{z_n^{s_n}}f,g)\ ,$$

and so

$$(*)\qquad \int_S|z_1|^{2s_1}\dots|z_n|^{2s_n}\cdot Xf\cdot\bar{g}dm=\int_S z_1^{s_1}\dots z_n^{s_n}Xf\cdot\overline{z_1^{s_1}\dots z_n^{s_n}g}dm\ .$$

Denote by $\mathcal{U}_{z^s}$, T_{z^s} the operators $\mathcal{U}_{z_1^{s_1}\dots z_n^{s_n}}$, $T_{z_1^{s_1}\dots z_n^{s_n}}$ respectively. Let $k\geq 0$ be an arbitrary integer number. Let $c_{ks}=$ $=\frac{k!}{s_1!\dots s_n!}$, where $s_1+\dots+s_n=|s|=k$. Then by (*) we have

$$\Big(\sum_{|s|=k}c_{ks}T^*_{z^s}T_{z^s}Xf,g\Big)=\int_S(|z_1|^2+\dots+|z_n|^2)^kXf\cdot\bar{g}dm=(Xf,g),$$

for $f\in H^2(\mu)$, $g\in H^2$. Hence

$$Xf=\sum_{|s|=k}c_{ks}T^*_{z^s}T_{z^s}Xf=\sum_{|s|=k}c_{ks}T^*_{z^s}X\mathcal{U}_{z^s}f\ .$$

It follows that

$$|(Xf,g)|\le\sum_s c_{ks}|(X\mathcal{U}_{z^s}f,T_{z^s}g)|\le$$

$$\le\sum_s\|c_{ks}^{\frac12}X\mathcal{U}_{z^s}f\|_m\cdot\|c_{ks}^{\frac12}T_{z^s}g\|_m\le$$

$$\le\sum_s c_{ks}\|\mathcal{U}_{z^s}f\|_\mu\cdot\|T_{z^s}g\|_m\le$$

$$\le[\sum_s c_{ks}\|\mathcal{U}_{z^s}f\|_\mu^2]^{\frac12}[\sum_s c_{ks}\|T_{z^s}g\|_m^2]^{\frac12}=$$

$$=[\int_B\sum_s c_{ks}|z_1|^{2s_1}\ldots|z_n|^{2s_n}|f|^2d\mu]^{\frac12}[\int_S(|z_1|^2+\ldots+|z_n|^2)^k|g|^2dm]^{\frac12}=$$

$$=[\int_B(|z_1|^2+\ldots+|z_n|^2)^k|f|^2d\mu+\int_S|f|^2d\mu]^{\frac12}\|g\|_m\ .$$

Now letting $k\to\infty$ we have by the Lebesgue theorem

$$|(Xf,g)|\le[\int_S|f|^2d\mu]^{\frac12}\cdot\|g\|_m=A_f\cdot\|g\|_m\ .$$

Thus $\|Xf\|\le A_f$ and so for any $p\in\mathcal{P}$ we have

$$\|Xp\|_m^2=\int_S|p\psi|^2dm\le\int_S|p|^2dm\ .$$

Hence (as before) for any positive, continuous function v on S

$$\int_S v|\psi|^2dm\le\int_S vd\mu$$

and so $\frac{d\mu_o}{dm}\ge|\psi|^2$, a.e. [m]. The proof is complete.

Applying Propositions 1 and 2 we have

PROPOSITION 3. *Let μ be a finite, positive Borel measure in C^n with compact support. Then $\mathcal{U}_\mu\sim\sim T$ if and only if*

(a) *there is a cyclic function $\psi\in H^2$ such that $\frac{d\mu_o}{dm}\ge|\psi|^2$ a.e. [m];*

(b) *there is a cyclic function $\phi\in H^2(\mu)$ such that*

$$\int |p|^2 |\phi|^2 d\mu \leq \int_S |p|^2 dm \;, \quad \textit{for every } p\varepsilon P \;.$$

PROOF. Assume that $U_\mu \sim\sim T$. By Proposition 1 of [3] we have $(\text{supp } \mu)^\wedge = (\text{supp } m)^\wedge = \overline{B}$, where $\wedge$ denotes the polynomially convex hull. Thus $\text{supp}\mu \subseteq \overline{B}$. Applying Propositions 1 and 2 we obtain (a) and (b).

Conversely, suppose that (a) and (b) hold true. Then for any integer $k \geq 0$

$$\int (|z_1|^2 + \ldots + |z_n|^2)^k |\phi|^2 d\mu \leq m(S) \;.$$

Since ϕ is cyclic (this is possible only if $\int_{C^n \setminus B} |\phi|^2 d\mu = 0$) and so $\text{supp}\,\mu$ must be contained in $\overline{B}$. Now we define bounded operators $X: H^2(\mu) \to H^2$, $Y: H^2 \to H^2(\mu)$ by $Xp = \psi p$, $Yp = \phi p$, $p\varepsilon P$. By (a) and (b) we have $||X|| \leq 1$ and $||Y|| \leq 1$. Since ϕ and ψ are cyclic it follows that X and Y have dense ranges. On the other hand since ψ is cyclic we also have $m << \mu_o$. It follows that Ker X=0. A similar reasonning proves that Ker Y=0. The proof is complete.

REMARK. In the above proof we have used the equality $(\text{supp } \mu)^\wedge = (\text{supp } m)^\wedge$, which holds for $U_\mu \sim\sim T$. It turns out that for any jointly subnormal n-tuples of commuting operators $S = (S_1, \ldots, S_n)$ and $S' = (S_1', \ldots, S_n')$, the assumption $S \sim\sim S'$ implies $\widehat{\sigma(S)} = \widehat{\sigma(S')}$, where $\sigma(S)$ (resp. $\sigma(S')$) is the Taylor joint spectrum of S (resp. S'), see [5].

REMARKS

Now we shall make some comments concerning conditions (a) and (b) of Proposition 3.

We start with condition (a). We want to know how to construct cyclic functions in H^2. For simplicity assume that n=2, but the method given below works also for n>2.

Let

$$A^2 = \{f \mid f \text{ holomorphic in the unit disc } D \text{ and } \int_D |f|^2 dxdy < +\infty\}$$

be the Bergman space.

If $f,g\varepsilon A^2$, then it is easy to check that $f\cdot g\varepsilon H^2$. Moreover, if $f,g\varepsilon A^2$ are both cyclic in A^2, then $f\cdot g$ is cyclic in H^2. In fact, denoting by $||\cdot||$ the norm in A^2, there exist sequences p_k, w_k of polynomials such that $||1-p_k f|| \to 0$ and $||1-w_k g|| \to 0$, as $k \to \infty$. Hence

$$\left(\int_S |1-p_k(z_1)w_k(z_2)f(z_1)g(z_2)|^2 dm\right)^{\frac{1}{2}} \leq \sqrt{2}\,(\pi||1-p_k f||+||p_k f||\cdot||1-w_k g||) .$$

But $||p_k f||<M$, $k=1,2,\ldots$, so the right hand side of the above inequality tends to zero as $k \to \infty$. Since there are known sufficient conditions for $f\varepsilon A^2$ to become a cyclic vector in A^2, see [7], one obtains a simple method to produce cyclic vectors in H^2. Thus one can give easily examples of measures satisfying condition (a).

Now let us consider condition (b). For which measures μ does there exist a cyclic vector $\phi\varepsilon H^2(\mu)$ and a constant $C>0$ such that

$$(\alpha) \qquad \int_{B_n} |p|^2|\phi|^2 d\mu \leq C\int_S |p|^2 dm, \text{ for every } p\varepsilon P \text{ ?}$$

Put $\phi=1$. Then by (α) we have

$$\frac{d\mu_o}{dm} \leq C , \qquad \text{and}$$

$$(**) \qquad \int_B |p|^2 d\mu \leq C\int_S |p|^2 dm , \qquad p\varepsilon P .$$

Therefore we have to see when (**) is satisfied. But the answer was given by Hörmander in [2]. We describe briefly his solution. For $x\varepsilon S$ we denote

$$\Pi_x = \{w\varepsilon \mathbb{C}^n, \ \sum_{i=1}^n w_i\bar{x}_i = 1\} .$$

Let

$$K(x,\sqrt{t}) = \{u\varepsilon \Pi_x , \ \sum_{i=1}^n |u_i - x_i|^2 < t, \ t>0\} .$$

Define

$$A_{(x,t)}=\{z\varepsilon\mathbb{C}^n, \ \forall y\varepsilon K(x,\sqrt{t}), \ \text{dist } (y,z)\le t\} \ .$$

Then by Theorem 4.3 of [2] condition (**) holds, if there exists $M>0$ such that

$$\mu(A_{(x,t)})\le Mt^n, \qquad \forall x\varepsilon S, \ \forall t>0 \ .$$

REMARK 1. Note that any function $\rho\neq 0$ bounded and holomorphic in B and such that $1/\rho\varepsilon H^2(\mu)$, is also cyclic for U_μ. Indeed, let $p_k\varepsilon P$ and

$$||1/\rho-p_k||_\mu \to 0, \qquad \text{as } k\to\infty \ .$$

Then

$$||1-p_k\rho||_\mu\le||\rho||_\infty||1/\rho-p_k||_\mu \to 0, \qquad \text{as } k\to\infty \ .$$

Therefore we can take $\phi=\rho$ and repeat the above reasonning for the measure $d\mu'=|\rho|^2 d\mu$.

REMARK 2. If $\int_B (1-|\xi|^2)^{-n}d\mu<+\infty$, then (**) holds true. In fact, for any $p\varepsilon P$ and $\xi\varepsilon B$ we have

$$|p(\xi)|^2\le||p||_m^2(1-|\xi|^2)^{-n}c_n \ .$$

It follows that (**) holds with

$$C=c_n\int_B(1-|\xi|^2)^{-n}d\mu \ .$$

REFERENCES

1. Clary, W.S.: Quasisimilarity of subnormal operators, *Ph.D. Thesis*, Univ. of Michigan, 1974.

2. Hörmander, L.: L^p-estimates for (pluri) subharmonic functions, *Math.Scand.* 20(1967), 68-78.

3. Hastings, W.W.: Commuting subnormal operators simultaneously quasisimilar to unilateral shifts, *Illinois J.Math.* 22(1978), 506-519.

4. Janas, J.: Lifting of commutant of subnormal operators and spectral inclusion theorem, *Bull.Acad.Polon.Sci.* 26(1978), 513-520.

5. Janas, J.: Remarks on similarity and quasisimilarity of operators, preprint, 1981.

6. Mlak, W.: Commutantss of subnormal operators, *Bull.Acad.Polon. Sci.* 9(1971), 837-842.

7. Shapiro, H.: Some remarks on weight polynomial approximation of holomorphic functions, *Mat.Sb.* 73(1967), 320-330.

J.Janas
Institute of Mathematics,
Polish Academy of Sciences,
ul. Solskiego 30, 31-027 Krakow,
Poland.

SOME PROPERTIES OF MASA'S IN FACTORS

V.Jones and S.Popa

§1. INTRODUCTION

In a recent paper ([14]) the second author showed that if N is a subfactor of a finite factor M with trivial relative commutant, one may find a MASA (maximal abelian *-subalgebra) A in N which is also MASA in M. It was also possible to control certain unitaries normalizing A and this led to some further questions. In this paper we give some answers to these questions.

The first paper to exhibit different kinds of MASA's in II_1 factors was that of Dixmier, [5]. If $A \subset M$ is a MASA he defined the normalizer $N(A)$ to be the set of those unitaries u in M such that $uAu^* = A$. A MASA is then said to be

a) regular if $N(A)'' = M$;

b) singular if $N(A)'' = A$;

c) semiregular if $N(A)''$ is a factor.

Dixmier gave examples of all three kinds of behaviour.

Another property for MASA's was first considered by Ambrose and Singer. Let M act in the standard form on $L^2(M,\tau)$, τ being the trace of M. Then the involution J satisfies $JMJ=M'$. A MASA A is said to be

d) simple if $(A \cup JAJ)''$ is a MASA in the algebra of all bounded operators on $L^2(M,\tau)$.

A question that arises naturally is the hereditarity of these properties, i.e. if A has one of a), b), c) or d) with respect to M and N is a subfactor between A and M, then does A have that property with respect to N? The hereditary property for a) has been proved by H.Dye in [6] (see also [8] for more general results). We give in Section 2 a proof that is slightly different from Dye's. Property b) is clearly hereditary and d) is also easily seen to be hereditary. But we give an example when A is semiregular in M and singular in some subfactor N containing A, so that c) is not hereditary.

In [14] it is shown that all II_1 factors have semiregular MASA's. Moreover if N is a subfactor of M with trivial relative commutant then there is a MASA A of M contained in N and semiregular in N. If N is hyperfinite then A may be chosen regular in N.

This immediately gives rise to a further question. If N and M are as above, when can A be chosen to be regular in M? As it is not known whether any II_1 factor has a regular MASA, this question is most interesting when M is hyperfinite. We show in Section 3 that in this case it is sufficient to suppose that N itself is regular as a subfactor. We conjecture that it is also necessary that N contain a regular subfactor. We give examples of subfactors which do not contain regular MASA's of M. These examples do not contain simple MASA's either.

In Section 4 we show that given any regular subfactor R_o of the hyperfinite II_1 factor R, with trivial relative commutant in R, there are two MASA's of R contained in R_o with normalizers generating R_o, which are not conjugate in R. This shows that there are uncountably many nonconjugate semiregular MASA's in R. It also shows that conjugacy of MASA's (even of regular ones) in a subfactor (even a regular one) does not imply conjugacy in the whole factor.

We want to mention that in Section 3 and 4 we heavily use Ocneanu's theorem on outer conjugacy of actions of countable amenable groups on R ([12]) and the important results of A.Connes in [2].

§2. REGULARITY IS A HEREDITARY PROPERTY WHILE SEMIREGULARITY IS NOT

The following three results are due to H.Dye ([6], Section 6; see also [8]). We give here more constructive proofs. They lead up to the fact that regularity is hereditary.

2.1. LEMMA. *Let* N *be a finite von Neumann algebra and let* $A \subset N$ *be a* MASA *in* N. *Suppose* Φ *is an automorphism of* A *and* $v \in N$

is a partial isometry such that

$$(*) \qquad va=\Phi(a)v, \qquad \textit{for all } a \textit{ in } A.$$

Then v *may be extended to a unitary* u *in the normalizer of* A *in* N.

PROOF. Denote by $e=v^*v$, $f=vv^*$ the left and respectively the right support of v in N. Since $v^*va=v^*\Phi(a)v=(\Phi(a^*)v)^*v=(va^*)^*v= =av^*v$ for all a in A, it follows that $e=v^*v\in A'\cap N=A$. Similary $f=vv^*\in A$. If we put $a=e$ and $a=\Phi^{-1}(f)$ in (*), then we get $ve=\Phi(e)v$ and $v\Phi^{-1}(f)=fv$ so that $\Phi(e)\geq f$ and $\Phi^{-1}(f)\geq e$, that is $\Phi(e)=f$.

To define the unitary u it is enough to find a partial isometry w, from e-ef onto f-ef, which implements an isomorphism of $A(e-ef)$ onto $A(f-ef)$ ($u=v+w^*+(1-e\ f)$ will then satisfy all the conditions).

Since $vA=Av$ it follows that $vAv^*\subset A$ so that $v^nAv^{*n}\subset A$, $n\geq 1$, and thus v^n and $(f-ef)v^n(e-ef)$ are partial isometries for all $n\geq 1$ and the left and right supports of these isometries are in A. Let e_n and f_n be the right and left supports respectively of $w_n=(f-ef)v^n(e-ef)$. Then the map $a\to w_naw_n^*\in Af_n$ for $a\in Ae_n$ is an isomorphism, for every $n\geq 1$. Moreover $\{e_n\}_{n\geq 1}$ (resp. $\{f_n\}_{n\geq 1}$) are mutually orthogonal projections such that $\sum_{n\geq 1} e_n=e-ef$ (resp. $\sum_{n\geq 1} f_n=f-ef$). Indeed, if $n>m$ then

$$v^nv^{*m}(f-ef)=v^{n-m}(v^mv^{*m}(f-ef))=v^{n-m}(f-ef)v^mv^{*m}=0$$

since $v(f-ef)=0$. It follows that

$$v^{*n}(f-ef)v^nv^{*m}(f-ef)v^m=0$$

and

$$e_ne_m=(e-ef)v^{*n}(f-ef)v^nv^{*m}(f-ef)v^m=0.$$

Similary $f_nf_m=0$ and

$$v^n(e-ef)v^{*n}v^m(e-ef)v^{*m}=0.$$

Let $e_o=(e-ef)-\sum_{n\geq 1} e_n$. Then the projections $v^ne_ov^{*n}$, $n\geq 1$ are mutually orthogonal since $v^ne_ov^{*n}\leq v^n(e-ef)v^{*n}$, $n\geq 1$. Since e_o is orthogonal to e_n, then $(f-ef)v^ne_o=0$ (by the definition of e_n), so $(f-ef)v^ne_ov^{*n}=0$. By induction it follows that $v^ne_ov^{*n}\leq e$ and that all the projections $v^ne_ov^{*n}$ are equivalent. Since N is finite this is impossible unless $e_o=0$. Thus $\sum_{n\geq 1} e_n=e-ef$ and similary

$\sum_{n\geq 1} f_n = f-ef$. Letting $w = \sum_{n\geq 1} w_n$ the proof is completed. Q.E.D.

Let M be a finite von Neumann algebra. If $N\subset M$ is a von Neumann subalgebra then there exists a normal conditional expectation onto N. Moreover if $N'\cap M\subset N$ then the conditional expectation is unique, faithful and it preserves the central trace of M (see [18]). This conditional expectation will be always denoted by E_N. In particular if N contains a MASA of M, then $N'\cap M\subset N$.

2.2. LEMMA. *Let* M *be a finite von Neumann algebra,* $A\subset M$ *a* MASA *in* M *and* N *a von Neumann subalgebra of* M *containing* A. *If* u *is a unitary in the normalizer* $\mathcal{N}(A)$ *of* A *in* M *then* $E_N(u)$ *is a partial isometry and there exists a unitary* v *in* $\mathcal{N}(A)\cap N$ *and a projection* e *in* A *such that* $E_N(u)=ve=ue$. *Moreover* e *is the maximal projection in* A *such that* $ue\in N$.

PROOF. Denote by Φ the automorphism of A implemented by u, $\Phi(a)=uau^*$ for all $a\in A$. Then $ua=\Phi(a)u$ and applying the conditional expectation E_N we get

$$E_N(u)a=\Phi(a)E_N(u), \qquad a\in A.$$

It follows that $E_N(u)^*E_N(u)\in A'\cap N=A$ so that $b=|E_N(u)|\in A$ and if e denotes the support of b then $e\in A$. Let w be the partial isometry in the polar decomposition of $E_N(u)$: $E_N(u)=wb$, $w^*w=e$. Then $wa=\Phi(a)w$ for all $a\in A$. We obtain that

$$u^*wa=u^*\Phi(a)w=au^*w$$

for all a in A, so that $u^*w\in A'\cap M=A$. Consequently

$$u^*w=E_N(u^*w)=E_N(u)^*w=bw^*w=b.$$

Since $b=u^*w$ is a partial isometry in A it follows that b is a projection, hence $b=e$ and $w=ue$. If $e_o\in A$ is a projection such that $ue_o\in N$ and $e_oe=0$, then

$$u(e+e_o)=E_N(u(e+e_o))=E_N(u)(e+e_o)=w(e+e_o)=w,$$

so that $e_o=0$ and e is the maximal projection in A with the property that $ue\in N$.

Using the finiteness of N, by Lemma 2.1 it follows that there exists a unitary v in $\mathcal{N}(A)\cap N$ such that $w=ve$. Q.E.D.

2.3. COROLLARY. *Let* M *be a finite von Neumann algebra and let* $A\subset M$ *be a regular* MASA *in* M. *If* N *is any von Neumann subalgebra of* M *containing* A *then* A *is a regular* MASA *in* N.

PROOF. By Lemma 2.2, $E_N(N(A)) \subset (N(A)\cap N)A$. Since $N(A)\cap N$ and A are contained in $(N(A)\cap N)''$ it follows that

$$N=E_N(M)=E_N(\overline{sp}^w N(A)) \subset (N(A)\cap N)''$$

so that $(N(A)\cap N)''=N$. Q.E.D.

2.4. REMARK. (see also [8]). Let M be a von Neumann algebra and let $N_o \subset M$ be a subalgebra in M such that $N_o'\cap M=Z(N_o)$ (or equivalently $N_o'\cap M\subset N_o$). Define the normalizer of N_o in M to be $N(N_o)=\{u\in M \mid u \text{ unitary such that } uN_ou^*=N_o\}$. If N is a subalgebra of M containing N_o and if there exists a normal conditional expectation E_N of M onto N then $E_N(N(N_o)) \subset (N(N_o)\cap N)''$. In particular if N_o is regular in M (i.e. $N(N_o)''=M$ or equivalently $\overline{sp}^w N(N_o)=M$) then N_o is regular in N.

Indeed, let $u\in N(N_o)$ and denote by $\Phi:N_o\to N_o$ the automorphism of N_o implemented by u ($\Phi(x)=uxu^*$). Then $v=E_N(u)$ is a partial isometry such that $vx=\Phi(x)v$, $x\in N_o$ (see the first part of the proof of Lemma 2.2). Moreover $e=v^*v$, $f=vv^*\in N_o'\cap M=Z(N_o)$. Since $\Phi(N_o)=N_o$ and $\phi(Z(N_o))=Z(N_o)$ it follows that $\phi(x)=vxv^*$ is an isomorphism of N_oe onto N_of. Using Zorn's lemma and the commutativity of $Z(N_o)$ it follows that there exist mutually orthogonal projections $\{e_i\}_{i\in I}$, e_o in $Z(N_o)$ such that $e_o+\sum_i e_i=e$ and such that $e_i\Phi(e_i)=0$, $i\in I$, $\Phi(e_o)=e_o$. If $w_o=ve_o$, $w_i=ve_i$, $i\in I$, then it follows that $u_o=w_o+(1-e_o)$ and $u_i=w_i+w_i^*+(1-w_i^*w_i-w_iw_i^*)$, $i\in I$, are unitary operators in N such that $u_iN_ou_i^*=N_o$, $u_oN_ou_o^*=N_o$. Thus $u_o,u_i\in N(N_o)\cap N$ so that

$$v=v(e_o+\sum_i e_i)=ve_o+\sum_i ve_i=u_oe_o+\sum_i u_ie_i\in((N(N_o)\cap N)\cup A)''=(N(N_o)\cap N)''.$$

However Lemma 2.1, which has an independent interest, does not seem to be true in this generality.

This shows that regularity is a hereditary property for MASA's. It is obvious that singularity is also.

To see that simplicity is hereditary, let $A\subset M$ be a simple MASA and let $A\subset N\subset M$. Denote $H=L^2(M,\tau)$ and $H_1=L^2(N,\tau)$. Then the standard representation of N is given by restriction to H_1 and the corresponding conjugation is just $J|H_1$. Since $AH_1\subset H_1$, $JAJH_1\subset H_1$ it follows that the algebra generated by $A|H_1$ and $JAJ|H_1$, is just the restriction of $B=(A\cup JAJ)''\subset B(H)$ to the invariant subspace H_1

and thus it is maximal abelian in $B(H_1)$.

We now proceed to construct an example of a II_1 factor M, a subfactor N, a MASA A of M with $A\subset N$, and such that A is singular in N but semiregular in M. Thus semiregularity is not a hereditary property.

The next lemma is probably known but does not seem to be proved anywhere.

2.5. LEMMA. *Let* N *be a von Neumann algebra and let* $A\subset N$ *be a* MASA *in* N. *Suppose that* $\alpha: G\to \mathrm{Aut}(N)$ *is an action of the discrete group* G *on* N *such that* $\alpha_g(A)=A$ *for all* $g\in G$. *Then* A *is a* MASA *in* $N\times_\alpha G$ *if and only if all the automorphisms* $\mathrm{Ad}(v)\circ\alpha_g$ *for* $v\in N(A)$ *(the normalizer of* A *in* N*),* $g\in G\setminus\{e\}$*, act freely on* A.

PROOF. Let u_g, $g\in G$, be the unitaries in $M\times_\alpha G$ canonicaly implementing the automorphisms α_g on N. If $\mathrm{Ad}(v)\circ\alpha_g$ is the identity on Af for some projection $f\in A$, $f\neq 0$, and for some unitary $v\in N(A)$ then fvu_g commutes with A and $fvu_g\notin A$ so that A is not maximal abelian in $N\times_\alpha G$.

Conversely suppose $\mathrm{Ad}(v)\circ\alpha_g$ acts freely on A for all $g\neq e$, $v\in N(A)$. Let $x=\sum_{g\in G} x_g u_g \in M\times_\alpha G$ be such that x commutes with A. Thus

$$\sum_{g\in G} a x_g u_g = \sum_{g\in G} x_g u_g a u_g^* u_g = \sum_{g\in G} x_g \alpha_g(a) u_g$$

so that $ax_g = x_g\alpha_g(a)$ for all $a\in A$, $g\in G$. For $g=e$ we get $x_e\in A'\cap N$. If $g\neq e$ and $x_g\neq 0$, take v to be the partial isometry in the polar decomposition of x_g. It follows as in Lemma 2.1 that $av=v\alpha_g(a)$, $a\in A$, and also v^*v, $vv^*\in A$. Since α_g acts freely on A which is commutative, we can find a nonzero projection $f\in A$, $f\leq vv^*$ such that $f\alpha_g(f)=0$. If $w=fv$ then

$$aw=afv=fav=fv\alpha_g(a)=w\alpha_g(a), \qquad a\in A.$$

Moreover $\alpha_g(a)=w^*aw$ for all $a\in Af$. If we define $u=w+w^*+(1-w^*w-ww^*)$, then u is a unitary element in N and $uAu^*=A$, so that $u\in N(A)$. In addition $\mathrm{Ad}(u)\circ\alpha_g(a)=u\alpha_g(a)u^*=w\alpha_g(a)w^*=a$ for all $a\in Af$.

But this is a contradiction since $\mathrm{Ad}(v)\circ\alpha_g$ acts freely on A by hypothesis. Thus $x_g=0$ for all $g\neq e$ and $x=x_e\in A'\cap N=A$. Q.E.D.

2.6. EXAMPLE. Let N be a factor and let $A\subset N$ be a singular MASA. If α is an automorphism of N such that $\alpha(A)=A$ and if α is not the identity on A then α is properly outer on N (since A is singular). If $\alpha|A$ is free, then by Lemma 2.4, A is maximal abelian in the factor $M=N\times\alpha$. Since the normalizer of A in $M=N\times\alpha$ contains $A\times(\alpha|A)$, it follows that if $\alpha|A$ is ergodic then A is semiregular in M but singular in N.

To construct such an example take G to be the group of affine transformations over $\mathbb{Q}$, with only positive homotheties, and let N be the von Neumann algebra associated to the left regular representation λ of G, i.e. $N=\lambda(G)''$. Since G is amenable, N is the hyperfinite II_1 factor ([2]). If $T,H\subset G$ denote the subgroups of translations and of positive homotheties in G respectively then T is a normal subgroup of G and $G/T\cong H\cong\mathbb{Q}_+^*$. As it is shown in [5], $A=\lambda(H)''\subset N$ is a singular MASA. Define $\gamma:G\to\mathbb{T}$ to be the following character on G. Fix $\{t_n\}$ to be a sequence of irrational numbers in the unit interval, linearly independent over $\mathbb{Z}$. If $\{p_n\}$ denote the positive prime numbers then take $\gamma_o:\mathbb{Q}_+^*\to\mathbb{T}$ to be the unique morphism such that $\gamma_o(p_n)=\exp(2\pi it_n)$ for all $n\geq 1$. Then take $\gamma|H=\gamma_o$ and $\gamma|T=1$.

Now define $\alpha_\gamma:N\to N$ to be the unique automorphism such that $\alpha_\gamma(\lambda(g))=\gamma(g)\lambda(g)$ for all $g\in G$ (see [13]). Then clearly $\alpha_\gamma(A)=A$ and α_γ is ergodic on A. Thus A is a semiregular MASA in the II_1 factor $M=N\times\alpha_\gamma$ although A is singular in the subfactor N. Note that by [2] M is also a hyperfinite factor.

§3. WHEN CAN A BE REGULAR IN M?

Let us begin by giving a sufficient condition in the case of the hyperfinite II_1 factor R.

3.1. THEOREM. *Suppose* $R_o\subset R$ *is a subfactor with trivial relative commutant. If* R_o *is regular in* R *(i.e. the normalizer* $N(R_o)$ *of* R_o *in* R *generates* R*) then there is a regular* MASA A *of* R *with* $A\subset R_o$.

PROOF. We begin by establishing that R is the crossed product of R_o by an outer action of some countable discrete group, as in [9]. The first thing to note is that two untaries in different cosets of $U(R_o)$ (the unitary group of R_o) in $N(R_o)$ are orthogonal. Indeed if u and v are such unitaries, u^*v induces an automorphism α of R_o, which is outer because $R_o' \cap R = \mathbb{C}$. Applying the conditional expectation E_{R_o} to the relation $u^*vx=\alpha(x)u^*v$ for $x \in R_o$, it follows that $E_{R_o}(u^*v)x=\alpha(x)E_{R_o}(u^*v)$, so that $E_{R_o}(u^*v)=0$.

Let $G=N(R_o)/U(R_o)$, a discrete group. By separability, G is countable. G is also amenable. To see this, choose for each $g \in G$ a $u_g \in N(R_o)$ representing g. Consider the vector spaces $\overline{u_g R_o}$, $g \in G$, which are mutually orthogonal in $L^2(R,\tau)$ (τ is the trace on R) and denote by P_g the orthogonal projection onto $\overline{u_g R_o}$. Let P be a conditional expectation from $B(L^2(R,\tau))$ onto R. Then if $f:G \to \mathbb{C}$ is an ℓ^∞ function, define

$$\mu(f)=\tau(P(\sum_{g \in G} f(g)P_g)).$$

Since $u_h P_g u_h^{-1}=P_{hg}$, μ is an invariant mean on $\ell^\infty(G)$. Hence G is amenable (see [17], p.210).

The u_g's satisfy $u_g u_h = w(g,h)u_{gh}$ for $w(g,h)$ a nonabelian, unitary two cocycle in R_o. By Connes' fundamental result ([2]), R_o is hyperfinite so we may apply Ocneanu's vanishing 2-cohomology result [12] to conclude that the u_g's may be modified so as to satisfy $u_g u_h = u_{gh}$. Then $g \to Ad(u_g)$ define an action α of G on R_o and the orthogonality of the u_g's ensures that R is in fact the crossed product $R_o \times_\alpha G$.

Now using the main result of [12] we may suppose the action α is any outer action we care to construct. So let us exhibit one with a globally invariant regular MASA. Let β be any free finite measure preserving action of G on a nonatomic probability space X (e.g. a Bernoulli shift action) and let $X^{\mathbb{Z}}$ be the probability space of all sequences $\{x_n\}$ in X indexed by $\mathbb{Z}$. The shift $x_n \to x_{n+1}$ induces an ergodic action of $\mathbb{Z}$ on $X^{\mathbb{Z}}$ which commutes with the action $g(\{x_n\})=\{\beta_g(x_n)\}$. We thus obtain a free ergodic measure preserving action of $G \times \mathbb{Z}$ on $A=L^\infty(X^{\mathbb{Z}},\mathbb{C})$. The crossed product $A \times (G \times \mathbb{Z})$ is a hyperfinite II_1 factor R with A as a regular MASA, and if $R_o = A \times \mathbb{Z} \subset R$ then the resulting action of G on R_o may be used as a model for the action α. Q.E.D.

3.2. REMARK. 1° The converse of Theorem 3.1 is not true: there are non-regular subfactors of R that contain regular MASA's. An example can be obtained by choosing an outer action of the finite group S_3 (i.e. the permutations of a set with three elements) on a hyperfinite II_1 factor M. Let $R=M\times S_3$ and $R_o=M\times S_2$, where S_2 is a subgroup of order 2 in S_3. Using Galois theory for factors one may check that $N(R_o)=U(R_o)$ but M itself contains a regular MASA of R by Theorem 3.1. Hence R_o does.

This leads immediately to the question: if R_o contains a regular MASA,A, is there a regular subfactor between R_o and A? The authors have been unable to answer this question, though one should note the structure theorem of Dye [6].

2°. The proof of Theorem 3.1 shows that the discrete group $N(R_o)/U(R_o)$ is a complete invariant for the conjugacy of regular subfactors R_o of R with trivial relative commutant (see [9]). A more complicated result is true for arbitrary regular subfactors.

We now make the following observation about finite group actions:

3.3. PROPOSITION. *If* M *is a factor and* G *is a finite group acting by outer automorphisms on* M *then* $(N(M^G))''=M^{[G,G]}$ *(here* M^G *denotes the fixed point algebra and* $[G,G]$ *the commutator subgroup).*

PROOF. Let M act in standard form on a Hilbert space and let $g\to u_g$ be the canonical unitary implementation of G. The algebra $(M'\cup\{u_g\})''$ is the commutant of M^G and it is naturally isomorphic to the crossed product of M' by the induced action of G (see [1]). If u is a unitary in $N(M^G)$ then Ad(u) induces an automorphisms of $(M^G)'$ which is the identity on M'. Now any such automorphism of the crossed product is given by $\Sigma a_g u_g \to \Sigma \chi(g) a_g u_g$ for some character $\chi: G\to\mathbb{T}$. Hence in particular $uu_g u^*=\chi(g)u_g$ and if $g\in[G,G]$, $u_g u u_g^*=u$. Thus $(N(M^G))''\subset M^{[G,G]}$.

To see the reverse inclusion note that any $\chi\in\mathrm{Char}(G)$ defines an automorphism of $(M^G)'$ as above, which will be implemented by a unitary u_χ in $N(M^G)$. So if $y'=\Sigma a_g' u_g \in N(M^G)'$ then $[u_\chi, y']=0$ so that $\Sigma\chi(g)a_g' u_g=\Sigma a_g' u_g$. Hence $a_g'=0$ for all $g\notin\ker\chi$. But $[G,G]=$ $= \bigcap_{\mathrm{Char}(G)} \ker\chi$. Hence $y'\in(M^{[G,G]})'$ and $N(M^G)'\subset(M^{[G,G]})'$. Q.E.D.

We now combine Theorem 3.1 and Proposition 3.3 to obtain:

3.4. THEOREM. *If* G *is a finite group of outer automorphisms of* R *then* R^G *contains a regular* MASA *if and only if* G *is abelian (*R *is as usual the hyperfinite* II_1 *factor).*

PROOF. Note first that the relative commutant of R^G in R is trivial. This well known fact follows immediately from the spatial situation as sketched in the proof of Proposition 3.3.

Now suppose G is abelian. Then $[G,G]=\{1\}$ so that R^G is regular by Proposition 3.3. Hence by Theorem 3.1 R^G contains a regular MASA.

Now suppose R^G contains a regular MASA A of R. To show that G is abelian it suffices to show that the group

$$\mathrm{Aut}_A R = \{\Phi \in \mathrm{Aut} R \mid \Phi|_A = \mathrm{id}_A\}$$

is abelian. This is well known but let us give a proof. If α and β are in $\mathrm{Aut}_A R$ and $u \in N(A)$ then $\alpha(uau^*)=uau^*$. Hence $u^*\alpha(u) \in A' \cap R = A$, so that $\alpha(u)=ua$ for some $a \in A$. Similary $\beta(u)=ub$ for $b \in A$. Thus

$$(\alpha\beta)(u)=\alpha(ub)=uab=(\beta\alpha)(u).$$

This shows that $(\alpha\beta)(u)=(\beta\alpha)(u)$ for all $u \in N(A)$ and by regularity $\alpha\beta=\beta\alpha$. Q.E.D.

For examples of finite nonabelian actions on R see [10].

The proof of the "$\Longrightarrow$" implication did not use any specific properties of R and holds for an arbitrary factor. We note a strengthening of this remark.

3.5. THEOREM. *If* G *is a finite nonabelian group of outer automorphisms of a factor* M *then* M^G *contains no simple* MASA.

PROOF. Let A be a MASA of M, contained in M^G. Put M in standard form with involution J, and let u_g be the canonical implementing unitaries. We have $Ju_gJ=u_g$ so that each u_g commutes with both A and JAJ. But if A is simple $(A \cup JAJ)''$ is a MASA on the Hilbert space and would thus contain the u_g's. This is impossible if G is nonabelian. Q.E.D.

This result gives another proof that M^G contains no regular MASA for G nonabelian, since regularity implies simplicity ([15]).

§4. TWO FAMILIES OF SEMIREGULAR MASA'S IN R

Let M be a type II_1 factor and $A \subset M$ a MASA. The existence of nontrivial central sequences for M in A ([2],[11]) is an invariant for A up to conjugacy by automorphisms of M. This fact was used by A.Connes and V.Jones in [3] to provide a separable II_1 factor with two nonconjugate regular MASA's. We use the techniques in the proof of Theorem 3.1 and some results from [16] in order to get the following:

4.1. THEOREM. *Let* $R_o \subset R$ *be a subfactor of* R, *with trivial relative commutant in* R *and which is regular in* R, *i.e.* $N(R_o)''=R$. *Then there exist two* MASA's *of* R, A_1 *and* A_2, *such that* $A_1, A_2 \subset R_o$, $N(A_1)''=N(A_2)''=R_o$ *and such that* A_2 *has nontrivial central sequences for* R *while* A_1 *has not.*

PROOF. As in Theorem 3.1, using [2] and [12] it follows that there exist a discrete countable amenable group G acting freely on R_o such that $R=R_o \times G$ and such that the inclusion of R_o in R becomes the natural inclusion of R_o in $R_o \times G$. Moreover by Ocneanu's theorem this identification does not depend on the action of G on R_o.

Consider a sequence $\{B_n\}_{0\leq n\leq \infty}$ of regular MASA's in R , which are mutually orthogonal in the sense of [16]. Such a sequence can be obtained as follows: let $u,v \in R$ be two unitaries generating R and satisfying $uv=\alpha vu$, for some $\alpha \in \mathbb{C}$, $\alpha^n \neq 1$ for all $n \in \mathbb{Z}$ (for instance consider R as obtained by the group measure construction for the irrational rotation of the one dimensional thorus). Let

$$B_o=\{u^n \mid n\in \mathbb{Z}\}'',$$

$$B_\infty=\{v^n \mid n\in \mathbb{Z}\}'',$$

$$B_k=\{(uv^k)^n \mid n\in \mathbb{Z}\}'', \qquad 1\leq k<\infty.$$

Then it is easy to see that B_k are regular MASA's of R and that, for $i\neq j$, B_j is orthogonal to B_i.

Let $\{B_g\}_{g\in G}$ be a relabelling of $\{B_k\}$ and denote by $\{R_g\}_{g\in G}$ a sequence of factors all isomorphic to R. Let $R_o=\bar{\otimes}_G R_g$, $A_1=\bar{\otimes}_G A_g$ and let $\alpha: G\to \mathrm{Aut}(R_o)$ be the Bernoulli shift action. Since tensor products preserve orthogonality it follows that $\alpha_g(A_1)$ is orthogonal to A_1 for all $g\neq e$. Let $R=R_o \times_\alpha G$ and denote by u_g the unitaries in R canonically implementing the automorphisms α_g. Thus $u_g A_1 u_g^*=\alpha_g(A_1)$ is orthogonal to A_1, for $g\neq e$, so that by Lemma 1.3

in [16] u_g is orthogonal to the normalizer $N(A_1)$ of A_1 in R. Consequently u_g is orthogonal to $N(A_1)$"and since $R_o \subset N(A_1)$" we get that $u_g R_o$ is orthogonal to $N(A_1)$" for all $g \neq e$, so that $N(A_1)"=R_o$.

Moreover if ω is a free ultrafilter on $\mathbb{N}$ and $\{e_n\} \subset A_1$ is a sequence of projections then $\tau(u_g e_n u_g^* e_n) = \tau(e_n)^2$, $n \in \mathbb{N}$, $g \in G$ (see [16], Lemma 1.1) while if $\{e_n\}$ is central in R then

$$\lim_{n\to\omega} \tau(u_g e_n u_g^* e_n) = \lim_{n\to\omega} \tau(e_n).$$

Thus

$$\lim_{n\to\omega}(\tau(e_n) - \tau(e_n)^2) = 0$$

so that $\{e_n\}$ is trivial and A_1 does not contain nontrivial central sequences of M.

To construct A_2 consider $R_o = (\bar{\otimes}_G R_g) \bar{\otimes} R^o$ where R^o is also isomorphic to R. Let $B \subset R^o$ be a regular MASA and denote by $A_2 = (\bar{\otimes}_G A_g) \bar{\otimes} \bar{\otimes} B$. Let G act on R_o as $\alpha \times \mathrm{id}$ and denote $R = R_o \times_{\alpha \times \mathrm{id}} G$. If v_g are the unitaries in R implementing the action $\alpha \times \mathrm{id}$ and if $A = (\bar{\otimes}_G B_g) \otimes 1$ then $v_g A v_g^*$ is orthogonal to A_2 so that, by Lemma 1.3 in [16], $v_g R_o$ is orthogonal to $N(A_2)$", for $g \neq e$, and thus $N(A_2)"=R_o$.

Finally, if we consider $\{x_n\} \subset 1 \otimes B \subset A_2$ to be any nontrivial bounded sequence which is central in $1 \otimes R^o$, then $\{x_n\}$ is also central in $R = R_o \times_{\alpha \times \mathrm{id}} G$, since it commutes with v_g and with $(\bar{\otimes} R_g) \otimes 1$. Q.E.D.

4.2. REMARK. In the proof of Theorem 4.1 one can use Theorem 9 in [7], instead of Lemma 1.3 in [16], to obtain that A_1, A_2 satisfy $N(A_1)"=N(A_2)"=R_o$.

4.3. COROLLARY. *There are uncountably many nonconjugate semiregular* MASA's *in* R.

PROOF. As pointed out in 3.2, 2°, there is a bijective correspondence between the classes of isomorphisms of discrete countable amenable groups and the conjugacy classes of regular subfactors of R with trivial relative commutant. By Theorem 4.1 every such subfactor R_o in R contains a MASA of R such that its normalizer in R generates R_o. Hence the corresponding group $G = N(R_o)/U(R_o)$ is also an invariant for this semiregular MASA. Since there exist continuously many nonisomorphic discrete countable amenable groups, the statement follows. Q.E.D.

4.4. COROLLARY. *Let* $R_o \subset R$ *be a regular subfactor,* $R_o \neq R$, *with trivial relative commutant. There are two regular* MASA's *of* R_o, *having the same normalizer in* R *as in* R_o, *which are not conjugated by automorphisms of* R.

(Note that by [4] any two regular MASA's in R_o are conjugated by an automorphism in R_o).

The authors would like to express their thanks to the Mathematics Institute of the University of Warwick where this work was completed.

REFERENCES

1. Aubert, P.L.: Théorie de Galois pour une W*-algèbre, *Comment.Math.Helvetici*, 39(51) (1976), 411-433.

2. Connes, A.: Classification of injective factors, *Ann.of Math.*, 104(1976), 73-115.

3. Connes, A.; Jones, V.: A II_1 factor with two nonconjugate Cartan subalgebras, *Bull. Amer.Math.Soc.*, to appear.

4. Connes, A.; Feldmann, J.; Weiss, B.: Amenable equivalence relations are generated by a single transformation, preprint.

5. Dixmier, J.: Sous-anneaux abéliens maximaux dans les facteurs de type fini, *Ann.of Math.*59(1954), 279-286.

6. Dye, H.: On groups of measure preserving transformations. II, *Amer.J.Math.*85(1963), 551-576.

7. Feldmann, J.; Moore, C.C.: Ergodic equivalence relations, cohomology and von Neumann algebras. II, *Trans.Amer.Math. Soc.*234(1977), 325-361.

8. Haga, Y.; Takeda, Z.: Correspondence between subgroups and subalgebras in a cross product von Neumann algebra, *Tôhoku Math.J.*24(1972), 167-190.

9. Jones, V.: Sur la conjugaison des sous-facteurs de type II_1, *C.R.Acad.Sci.Paris, Sér.A* 284(1977), 597-598.

10. Jones, V.: *Actions of finite groups on the hyperfinite type* II_1 *factor*, Memoirs A.M.S., no.237, 1980.

11. Mc Duff, D.: Central sequences and the hyperfinite factor, *Proc.London Math.Soc.* 21(1970), 443-461.

12. Ocneanu, A.: Actions des groupes moyennables sur les algèbres de von Neumann, *C.R. Acad.Sci.Paris, Sér.A*, 1980.

13. Paschke, W.L.: Inner product modules arising from compact automorphisms groups of von Neumann algebras, *Trans.Amer.*

*Math.Soc.*224(1976), 87-102.

14. Popa, S.: On a problem of R.V.Kadison on maximal abelian *-subalgebras in factors, *Invent.Math.*65(1981),269-281.

15. Popa, S.: On MASA's of algebras associated with free groups, *INCREST preprint*, no.40/1981.

16. Popa, S.: Orthogonal pairs of subalgebras in finite factors, *INCREST preprint*, no.89(1981).

17. Sakai, S.: *C*-algebras and W*-algebras*, Springer Verlag, 1971.

18. Strătilă, Ş.: *Modular theory*, Editura Academiei and Abacus Press, 1981.

19. Tauer, R.F.: Maximal abelian subalgebras in finite factors, *Trans.Amer.Math.Soc.*114(1965), 281-308.

V.R.F. Jones
Department of Mathematics,
University of California,
Los Angeles, CA 90024,
U.S.A.

Sorin Popa
Department of Mathematics,
INCREST,
Bdul Păcii 220,79622 Bucharest,
Romania.

GENERALIZED RESOLVENTS OF DUAL PAIRS OF CONTRACTIONS

Heinz Langer and Björn Textorius

1. Let H be a Hilbert space. The contractions T, S in H with closed, nondense domains $\mathcal{D}(T)$, $\mathcal{D}(S)$ are said to form a *dual pair* (denoted by $\{T,S\}$), if the relation

$$(Tx,y)=(x,Sy) \qquad (x\in\mathcal{D}(T),\ y\in\mathcal{D}(S))$$

holds. It is well-known that for each dual pair $\{T,S\}$ of contractions in H there exists a contraction $\tilde{T}$ in H, $\mathcal{D}(\tilde{T})=H$, such that

$$T\subset\tilde{T}, \quad S\subset\tilde{T}^*. \tag{1}$$

This follows from [11] (see also the proof of Theorem V.9.1 in [2]); the special case T=S goes back to M.G. Kreĭn [4] (see also [3], [10]). Lately Gr.Arsene and A.Gheondea [1] gave a description of all contractions $\tilde{T}$ in H which satisfy (1) for a given dual pair $\{T,S\}$ of contractions. (They even consider the more general situation that T and S act between two possibly different Hilbert spaces H and K:

$$T:\mathcal{D}(T)\to K, \quad S:\mathcal{D}(S)\to H$$

with $\mathcal{D}(T)\subset H$, $\mathcal{D}(S)\subset K$.)

In this note we introduce the generalized resolvents of a given dual pair $\{T,S\}$ of contractions in an analoguous way as they have been defined for unbounded Hermitian operators by M.G. Kreĭn and M.A.Neumark and for Hermitian contractions by M.G.Kreĭn and I.E.Ovčarenko [7]. The main result is a description of the set of all these generalized resolvents of $\{T,S\}$ (Theorems 1 and 2).

Theorem 1 below is an extension of the Theorem in [8] where the special case of a single nondensely defined contraction has been considered. At the same time, in Theorem 3 we give an extension of the Theorem in [8] for the case when instead of $\overset{\circ}{T}$ an arbitrary contraction extension of T is chosen. Moreover, it will be shown in §5 that the descriptions of the generalized resolvents of an isometric operator in Hilbert space, which follow from [9, Satz 2.6] and [5, Satz 4.1], are also immediate consequences of Theorem 3 (in [9], [5] the more general case of a Pontrjagin space $\Pi_\varkappa$ has been considered).

We mention that by means of the Cayley transformation the results of this note carry over to "dual pairs" of dissipative operators, see [2, V. 10], [12, IV. 4.2]. Thus a description of the "spectral functions" of a dual pair of dissipative operators can be obtained. This will be considered elsewhere.

2. Let $\{T,S\}$ be a dual pair of contractions in H. A contraction $\tilde{T}$ in some Hilbert space $\tilde{H} \supset H$, $\mathcal{D}(\tilde{T})=\tilde{H}$, with the properties (1), that is $\tilde{T}$ is an extension of T and $\tilde{T}^*$ is an extension of S, is said to be a *contraction extension* (c.e.) of $\{T,S\}$. If, in particular, $\tilde{H}=H$, the c.e. $\tilde{T}$ will be called *canonical*.

Let $\tilde{T}$ in $\tilde{H}$ be an arbitrary c.e. of the dual pair $\{T,S\}$, P_H the orthogonal projection from $\tilde{H}$ onto H. The operator function R_z:

$$z \to R_z := P_H(z\tilde{T}-\tilde{I})^{-1}|_H \qquad (|z|<1)$$

with values in $[H]$[1] is called a *generalized resolvent* (g.r.) of the dual pair $\{T,S\}$, or, sometimes, the g.r. of the dual pair $\{T,S\}$, generated by the c.e. $\tilde{T}$. If $\tilde{T}$ can be chosen canonical, then R_z is also called a *canonical g.r.* of $\{T,S\}$.

In [1] (comp. also [3]) it was shown that the contractions T,S of a dual pair $\{T,S\}$ admit the matrix representations[2]

$$(2) \qquad T=\binom{A}{\Gamma_2 D_A}:\mathcal{D}(T) \to \mathcal{D}(S) \oplus \mathcal{D}(S)^\perp,$$

$$(3) \qquad S=\binom{A^*}{\Gamma_1^* D_{A^*}}:\mathcal{D}(S) \to \mathcal{D}(T) \oplus \mathcal{D}(T)^\perp,$$

with some contractions $A\varepsilon[\mathcal{D}(T),\mathcal{D}(S)]$, $\Gamma_2\varepsilon[\mathcal{D}_A,\mathcal{D}(S)^\perp]$,

1) $[H,K]$ denotes the set of all bounded linear operators from all of H into K, and we put $[H]:=[H,H]$.

2) If $A\varepsilon[H,K]$ we define $D_A:=(I-A^*A)^{1/2}$ and $\mathcal{D}_A:=\mathcal{R}((I-A^*A)^{1/2})$; $\mathcal{R}(B)$ denotes the range of the linear operator B.

$\Gamma_1 \in [\mathcal{D}(T)^\perp, \mathcal{D}_{A^*}]$, and that there exists a 1,1-correspondence $G \to T_G$ between the set of all canonical c.e. T_G of $\{T,S\}$ and the set of all contractions $G \in [\mathcal{D}_{\Gamma_1}, \mathcal{D}_{\Gamma_2^*}]$ given by the matrix representation

$$(4) \qquad T_G = \begin{pmatrix} A & D_{A^*}\Gamma_1 \\ \Gamma_2 D_A & -\Gamma_2 A^* \Gamma_1 + D_{\Gamma_2^*} G D_{\Gamma_1} \end{pmatrix};$$

here the initial space is decomposed as $H = \mathcal{D}(T) \oplus \mathcal{D}(T)^\perp$ and the range space as $H = \mathcal{D}(S) \oplus \mathcal{D}(S)^\perp$.

PROPOSITION 1. (Comp. [1]). *For the canonical c.e.* T_G *in* (4) *the operator*

$$U = \begin{pmatrix} A & D_{A^*}\Gamma_1 & D_{A^*}D_{\Gamma_1^*} & 0 \\ \Gamma_2 D_A & -\Gamma_2 A^*\Gamma_1 + D_{\Gamma_2^*} G D_{\Gamma_1} & -\Gamma_2 A^* D_{\Gamma_1^*} - D_{\Gamma_2^*} G \Gamma_1^* & D_{\Gamma_2^*} D_{G^*} \\ D_{\Gamma_2} D_A & -D_{\Gamma_2} A^* \Gamma_1 - \Gamma_2^* G D_{\Gamma_1} & -D_{\Gamma_2} A^* D_{\Gamma_1^*} + \Gamma_2^* G \Gamma_1^* & -\Gamma_2^* D_{G^*} \\ 0 & D_G D_{\Gamma_1} & -D_G \Gamma_1^* & -G^* \end{pmatrix},$$

acting from $\mathcal{D}(T) \oplus \mathcal{D}(T)^\perp \oplus \mathcal{D}_{\Gamma_1^*} \oplus \mathcal{D}_{\Gamma_2^*}$ *into* $\mathcal{D}(S) \oplus \mathcal{D}(S)^\perp \oplus \mathcal{D}_{\Gamma_2} \oplus \mathcal{D}_{\Gamma_1}$ *is unitary.*[1)]

PROOF. If we decompose the matrix of U as

$$U = \begin{pmatrix} T_G & U_{12} \\ U_{21} & U_{22} \end{pmatrix} : H \oplus (\mathcal{D}_{\Gamma_1^*} \oplus \mathcal{D}_{\Gamma_2^*}) \to H \oplus (\mathcal{D}_{\Gamma_2} \oplus \mathcal{D}_{\Gamma_1}),$$

the statement is equivalent to the relations

$$(5) \qquad \begin{aligned} &T_G^* T_G + U_{21}^* U_{21} = I, && T_G T_G^* + U_{12} U_{12}^* = I, \\ &U_{12}^* U_{12} + U_{22}^* U_{22} = I, && U_{21} U_{21}^* + U_{22} U_{22}^* = I, \\ &U_{12}^* T_G + U_{22}^* U_{21} = 0, && T_G U_{21}^* + U_{12} U_{22}^* = 0, \end{aligned}$$

---

1) $V \in [K_1, K_2]$ is said to be unitary if it is isometric and $\mathcal{D}(V) = K_1$, $\mathcal{R}(V) = K_2$.

which can be checked without difficulty.

If H, K are two Hilbert spaces, by $\mathsf{K}(H,K)$ we denote the set of all contractive analytic functions, defined on $D:=\{z:|z|<1\}$ and with values in $[H,K]$; $\mathsf{K}_o(H,K)$ is the subset of $\mathsf{K}(H,K)$ consisting of those elements of $\mathsf{K}(H,K)$ which are independent of z.

The following proposition is well-known, see e.g. [6, p.233]; it will be given for the convenience of the reader.

PROPOSITION 2. *If* H, H_1, K_1 *are Hilbert spaces and*

$$U=\begin{pmatrix} U_{11} & U_{12} \\ U_{21} & U_{22} \end{pmatrix} : H \oplus H_1 \to H \oplus K_1$$

is a unitary operator, then the operator function Θ:

$$\Theta(z):=U_{22}-zU_{21}(zU_{11}-I)^{-1}U_{12} \qquad (|z|<1)$$

belongs to $\mathsf{K}(H_1,K_1)$.

Indeed, observing the relations (5) with $T_G=U_{11}$, it follows for $|z|<1$

$$\frac{I-\Theta(z)^*\Theta(z)}{1-|z|^2}=U_{12}^*(\bar{z}U_{11}^*-I)^{-1}(zU_{11}-I)^{-1}U_{12} .$$

Thus the operators $\Theta(z)$, $|z|<1$, are contractions, and the proposition is proved.

In the following, with the given dual pair $\{T,S\}$ of contractions in H, the operators A, Γ_1, Γ_2 and T_G are always defined according to (2), (3) and (4).

COROLLARY 1. *For an arbitrary contraction* $G\varepsilon[D_{\Gamma_1},D_{\Gamma_2^*}]$ *the operator function* Θ_G:

$$(6) \qquad \Theta_G(z):=\begin{pmatrix} -D_{\Gamma_2}A^*D_{\Gamma_1}+\Gamma_2^*G\Gamma_1^* & -\Gamma_2^*D_{G^*} \\ -D_G\Gamma_1^* & -G^* \end{pmatrix} -$$

$$-z\cdot\begin{pmatrix} D_{\Gamma_2}D_A & -D_{\Gamma_2}A^*\Gamma_1-\Gamma_2^*GD_{\Gamma_1} \\ 0 & D_GD_{\Gamma_1}\end{pmatrix}(zT_G-I)^{-1}\begin{pmatrix} D_{A^*}D_{\Gamma_1^*} & 0 \\ -\Gamma_2A^*D_{\Gamma_1^*}+\Gamma_2^*G\Gamma_1^* & D_{\Gamma_2^*}D_{G^*}\end{pmatrix}$$

belongs to $\mathrm{K}(\mathcal{D}_{\Gamma_1^*}\oplus\mathcal{D}_{\Gamma_2^*},\ \mathcal{D}_{\Gamma_2}\oplus\mathcal{D}_{\Gamma_1})$.

We mention that Θ_G is a characteristic function (in M.G. Kreĭn's sense, see [6]) of the operator T_G^*.

If H_o is a subspace of H, by P_{H_o} we denote the orthogonal projection onto H_o.

COROLLARY 2. *For an arbitrary contraction* $G\varepsilon[\mathcal{D}_{\Gamma_1},\mathcal{D}_{\Gamma_2^*}]$ *the operator function* X_G:

$$X_G(z):=P_{\mathcal{D}_{\Gamma_1}}\Theta_G(z)P_{\mathcal{D}_{\Gamma_2^*}}\qquad(|z|<1)\tag{7}$$

belongs to $\mathrm{K}(\mathcal{D}_{\Gamma_2^*},\ \mathcal{D}_{\Gamma_1})$.

The relations (6) and (7) imply

$$X_G(z)=-G^*-zD_GD_{\Gamma_1}P_{\mathcal{D}(T)^\perp}(zT_G-I)^{-1}P_{\mathcal{D}(S)^\perp}D_{\Gamma_2^*}D_{G^*}\qquad(|z|<1).$$

Therefore $X_G(z)$ maps $\mathcal{D}_{G^*}$ into $\mathcal{D}_G$, thus it belongs also to $\mathrm{K}(\mathcal{D}_{G^*},\mathcal{D}_G)$.

3. In the following the canonical c.e. T_o of the dual pair $\{T,S\}$, corresponding to the operator $G=0$ plays a special role:

$$T_o:=\begin{pmatrix} A & D_{A^*}\Gamma_1 \\ \Gamma_2D_A & -\Gamma_2A^*\Gamma_1\end{pmatrix}.$$

Its resolvent will be denoted by $\overset{o}{R}_z$:

$$\overset{o}{R}_z:=(zT_o-I)^{-1}.$$

The corresponding function X_o is given by

$$(8)\quad X_o(z)=-zD_{\Gamma_1}P_{\mathcal{D}(T)^\perp}(zT_o-I)^{-1}P_{\mathcal{D}(S)^\perp}D_{\Gamma_2^*}\qquad (|z|<1)\ .$$

THEOREM 1. *Let* {T,S} *be a dual pair of contractions in the Hilbert space* H. *The formula*

$$(9)\quad R_z=\mathring{R}_z-z\mathring{R}_zD_{\Gamma_2^*}G(z)(I-X_o(z)G(z))^{-1}D_{\Gamma_1}P_{\mathcal{D}(T)^\perp}\mathring{R}_z\qquad (|z|<1)$$

establishes a 1,1-*correspondence between all g.r.* R_z *of* {T,S} *and all functions* $G\in\mathsf{K}(\mathcal{D}_{\Gamma_1},\mathcal{D}_{\Gamma_2^*})$. *The g.r.* R_z *is canonical if and only if* $G\in\mathsf{K}_o(\mathcal{D}_{\Gamma_1},\mathcal{D}_{\Gamma_2^*})$.

PROOF. (a) Let $\tilde{T}$ be a canonical c.e. of {T,S}. Then, according to the results of [1] quoted above, there exists a contraction $G\in[\mathcal{D}_{\Gamma_1},\mathcal{D}_{\Gamma_2^*}]$ such that

$$\tilde{T}=T_G=T_o+D_{\Gamma_2^*}GD_{\Gamma_1}P\ ;$$

here we have put $P:=P_{\mathcal{D}(T)^\perp}$. It follows

$$(10)\quad \begin{aligned}(z\tilde{T}-I)^{-1}&=(zT_o-I+zD_{\Gamma_2^*}GD_{\Gamma_1}P)^{-1}=\mathring{R}_z(I+zD_{\Gamma_2^*}GD_{\Gamma_1}P\mathring{R}_z)^{-1}=\\ &=\mathring{R}_z-\mathring{R}_z[I-(I+zD_{\Gamma_2^*}GD_{\Gamma_1}P\mathring{R}_z)^{-1}]=\\ &=\mathring{R}_z-z\mathring{R}_z(I+zD_{\Gamma_2^*}GD_{\Gamma_1}P\mathring{R}_z)^{-1}D_{\Gamma_2^*}GD_{\Gamma_1}P\mathring{R}_z=\\ &=\mathring{R}_z-z\mathring{R}_zD_{\Gamma_2^*}G(I+zD_{\Gamma_1}P\mathring{R}_zD_{\Gamma_2^*}G)^{-1}D_{\Gamma_1}P\mathring{R}_z\ .\end{aligned}$$

Observing the relation (8) it follows

$$(z\tilde{T}-I)^{-1}=\mathring{R}_z-z\mathring{R}_zD_{\Gamma_2^*}G(I-X_o(z)G)^{-1}D_{\Gamma_1}P\mathring{R}_z\ .$$

(b) Let now $\tilde{T}$ be an arbitrary c.e. of {T,S}. As in [8] it follows that for the generalized resolvent R_z generated by $\tilde{T}$ the operator R_z^{-1}, $|z|<1$, exists; $T(z):=\frac{1}{z}(R_z^{-1}+I)$, extended by continuity to $z=0$, is a contraction in H for all $|z|<1$ and $T(z)\supset T$.

Then $T(z)^*=\frac{1}{\bar{z}}((R_z^*)^{-1}+I)$ is also a contraction and for $x\varepsilon\mathcal{D}(S)$ we have

$$(T(z)^*-S)x=\frac{1}{\bar{z}}(R_z^{-1})^*(I+R_z^*-\bar{z}R_z^*S)x=$$

$$=\frac{1}{\bar{z}}(R_z^{-1})^*(\tilde{P}(\bar{z}\tilde{T}^*-I)^{-1}(\bar{z}\tilde{T}^*-\bar{z}S)x=0 \ .$$

That is, $T(z)$ is a canonical c.e. of $\{T,S\}$. Therefore the result of (a) can be applied to $T(z)$. Thus for some contraction $G(z)\varepsilon$ $\varepsilon[\mathcal{D}_{\Gamma_1},\mathcal{D}_{\Gamma_2^*}]$ we have

$$T(z)=T_o+D_{\Gamma_2^*}G(z)D_{\Gamma_1}P$$

and since $R_z=(zT(z)-I)^{-1}$, the representation formula

$$R_z=\overset{o}{R}_z-z\overset{o}{R}_zD_{\Gamma_2^*}G(z)(I-X_o(z)G(z))^{-1}D_{\Gamma_1}P\overset{o}{R}_z$$

holds true. Moreover, if $x=D_{\Gamma_1}y$ for some $y\varepsilon\mathcal{D}(T)^{\perp}$, then

$$G(z)x=D_{\Gamma_2^*}^{-1}(T(z)-T_o)y \qquad (|z|<1).$$

As $T(z)$ is holomorphic in D this yields $G\varepsilon K(\mathcal{D}_{\Gamma_1},\mathcal{D}_{\Gamma_2^*})$. Therefore an arbitrary g.r. of $\{T,S\}$ admits a representation (9).

(c) Now we turn to the inverse problem. Let a function $G\varepsilon$ $\varepsilon K(\mathcal{D}_{\Gamma_1},\mathcal{D}_{\Gamma_2^*})$ be given. According to [12, Proposition V.2.1] its domain $\mathcal{D}_{\Gamma_1}$ and range space $\mathcal{D}_{\Gamma_2^*}$ can be decomposed:

$$\mathcal{D}_{\Gamma_1}=\mathcal{D}'\oplus\mathcal{D}'' \ , \qquad \mathcal{D}_{\Gamma_2^*}=\mathcal{D}_*'\oplus\mathcal{D}_*'' \ ,$$

such that the function G'', $G''(z):=G(z)|\mathcal{D}''$ belongs to $K(\mathcal{D}'',\mathcal{D}_*'')$ and is purely contractive, whereas $G'(z):=G(z)|\mathcal{D}'$ is a unitary operator from $\mathcal{D}'$ onto $\mathcal{D}_*'$ which is independent of z,

$$G'(z)=G' \qquad (|z|<1).$$

The purely contractive analytic function $G''\varepsilon K(\mathcal{D}'',\mathcal{D}_*'')$ is the characteristic function of some contraction $\hat{S}$ in a Hilbert space $\mathcal{H}_1$

(see [12, Theorem VI. 3.1]):

$$G''(z)=-\hat{S}-zD_{\hat{S}^*}(z\hat{S}^*-I)^{-1}D_{\hat{S}}:\mathcal{D}''\to\mathcal{D}'';$$

here $\mathcal{D}''$ and $\mathcal{D}''_*$ can be identified with $\mathcal{D}_{\hat{S}}$ and $\mathcal{D}_{\hat{S}^*}$ respectively. If we denote the orthogonal projections from $\mathcal{D}_{\Gamma_1}$ onto $\mathcal{D}'$ and $\mathcal{D}''$ by P' and P'' and those from $\mathcal{D}_{\Gamma_2^*}$ onto $\mathcal{D}'_*$ and $\mathcal{D}''_*$ by P'_* and P''_* respectively, the function G has finally the following representation:

$$G(z)=P''_*(-\hat{S}-zD_{\hat{S}^*}(z\hat{S}^*-I)^{-1}D_{\hat{S}})P''-P'_*G'P' .$$

Later we shall also make use of the relations

$$(I-P'')\hat{S}^*P''_*=0, \qquad (I-P''_*)\hat{S}P''=0 . \tag{11}$$

Indeed, with the identification made above the first (second) relation is a consequence of

$$\hat{S}^*\mathcal{D}_{\hat{S}^*}\subset\mathcal{D}_{\hat{S}} \qquad (\hat{S}\mathcal{D}_{\hat{S}}\subset\mathcal{D}_{\hat{S}^*} \text{ respectively}).$$

With the contraction $\tilde{S}\in[\mathcal{D}_{\Gamma_1},\mathcal{D}_{\Gamma_2^*}]$:

$$\tilde{S}:=P'_*G'P'+P''_*\hat{S}P''$$

we form the operator

$$\tilde{T}:=\begin{pmatrix} A & D_{A^*}\Gamma_1 & 0 \\ \Gamma_2 D_A & -\Gamma_2 A^*\Gamma_1-D_{\Gamma_2^*}\tilde{S}D_{\Gamma_1} & D_{\Gamma_2^*}P''_*D_{\hat{S}^*} \\ 0 & D_{\hat{S}}P''D_{\Gamma_1} & \hat{S}^* \end{pmatrix},$$

acting from $\tilde{H}:=\mathcal{D}(T)\oplus\mathcal{D}(T)^{\perp}\oplus H_1$ into $\tilde{H}=\mathcal{D}(S)\oplus\mathcal{D}(S)^{\perp}\oplus H_1$. Then $\tilde{T}$ is a contraction. Indeed, if we denote the operators in the corresponding matrix representation of $I-\tilde{T}^*\tilde{T}$ by B_{jk}, $j,k=1,2,3$, it follows

$$B_{11}=I-A^*A-D_A\Gamma_2^*\Gamma_2D_A=D_AD^2_{\Gamma_2}D_A ,$$

$$B_{12}=-A^*D_{A^*}\Gamma_1+D_A\Gamma_2^*(\Gamma_2A^*\Gamma_1+D_{\Gamma_2^*}\tilde{S}D_{\Gamma_1})=$$

$$=D_A(-A^*\Gamma_1+\Gamma_2^*\Gamma_2A^*\Gamma_1+\Gamma_2^*D_{\Gamma_2^*}\tilde{S}D_{\Gamma_1})=$$

$$=D_AD_{\Gamma_2}(-D_{\Gamma_2}A^*\Gamma_1+\Gamma_2^*\tilde{S}D_{\Gamma_1}) \ ,$$

$$B_{13}=-D_A\Gamma_2^*D_{\Gamma_2^*}P_*''D_{\hat{S}^*}=-D_AD_{\Gamma_2}\Gamma_2^*D_{\hat{S}^*} \ ,$$

$$B_{22}=I-\Gamma_1^*D_{A^*}^2\Gamma_1-(\Gamma_1^*A\Gamma_2^*+D_{\Gamma_1}\tilde{S}^*D_{\Gamma_2^*})(\Gamma_2A^*\Gamma_1+D_{\Gamma_2^*}\tilde{S}D_{\Gamma_1})-D_{\Gamma_1}D_{\hat{S}}^2P''D_{\Gamma_1}=$$

$$=(-\Gamma_1^*AD_{\Gamma_2}+D_{\Gamma_1}\tilde{S}^*T_2)(-D_{\Gamma_2}A^*\Gamma_1+\Gamma_2^*\tilde{S}D_{\Gamma_1})$$

(here we have used the relation

$$D_{\Gamma_1}(I-\tilde{S}^*\tilde{S}-D_{\hat{S}}^2P'')D_{\Gamma_1}=0 \ ,$$

which follows from the definition of $\tilde{S}$ and (11)),

$$B_{23}=(\Gamma_1^*A\Gamma_2^*+D_{\Gamma_1}\tilde{S}^*D_{\Gamma_2^*})D_{\Gamma_2^*}P_*''D_{\hat{S}^*}-D_{\Gamma_1}P''D_{\hat{S}}\hat{S}^*=$$

$$=[\Gamma_1^*AD_{\Gamma_2}\Gamma_2^*+D_{\Gamma_1}(P'G'^*P_*'+P''\hat{S}^*P_*'')(I-\Gamma_2\Gamma_2^*)-D_{\Gamma_1}P''\hat{S}^*]D_{\hat{S}^*}=$$

$$=(\Gamma_1^*AD_{\Gamma_2}-D_{\Gamma_1}\tilde{S}^*\Gamma_2)\Gamma_2^*D_{\hat{S}^*}$$

(observe that $P_*'D_{\hat{S}^*}=0$) and

$$B_{33}=I-D_{\hat{S}^*}P_*''D_{\Gamma_2^*}^2P_*''D_{\hat{S}^*}-\hat{S}\hat{S}^*=D_{\hat{S}^*}(I-P_*''D_{\Gamma_2^*}^2P_*'')D_{\hat{S}^*}=$$

$$=D_{\hat{S}^*}P_*''(I-D_{\Gamma_2^*}^2)P_*''D_{\hat{S}^*}=D_{\hat{S}^*}P_*''\Gamma_2\Gamma_2^*P_*''D_{\hat{S}^*}.$$

Thus we get

$$I-\tilde{T}^*\tilde{T}=\begin{pmatrix} D_AD_{\Gamma_2} & 0 & 0 \\ 0 & I & 0 \\ 0 & 0 & D_{\hat{S}^*}P_*''\Gamma_2 \end{pmatrix}\cdot\begin{pmatrix} I & C & -I \\ C^* & C^*C & -C^* \\ -I & -C & I \end{pmatrix}\cdot\begin{pmatrix} D_{\Gamma_2}D_A & 0 & 0 \\ 0 & I & 0 \\ 0 & 0 & \Gamma_2^*P_*''D_{\hat{S}^*} \end{pmatrix}$$

with $C:=-D_{\Gamma_2}A^*\Gamma_1+\Gamma_2^*\tilde{S}D_{\Gamma_1}$. Observing that the operator

$$\begin{pmatrix} I & C & -I \\ C^* & C^*C & -C^* \\ -I & -C & I \end{pmatrix} = \begin{pmatrix} I \\ C^* \\ -I \end{pmatrix} \cdot (I \quad C \quad -I)$$

is nonnegative, it follows that $\tilde{T}$ is a contraction.

With respect to the decomposition $\tilde{H}=H \oplus H_1$ the operator $\tilde{T}$ can be written as

$$\tilde{T}=\begin{pmatrix} T_o-D_{\Gamma_2^*}\tilde{S}D_{\Gamma_1} & D_{\Gamma_2^*}P''_*D_{\hat{S}^*} \\ D_{\hat{S}}P''D_{\Gamma_1}P & \hat{S}^* \end{pmatrix}.$$

Hence

$$P_H(z\tilde{T}-\tilde{I})^{-1}|_H=(zT_o-I-zD_{\Gamma_2^*}\tilde{S}D_{\Gamma_1}P-z^2D_{\Gamma_2^*}P''_*D_{\hat{S}^*}(z\hat{S}^*-I)^{-1}D_{\hat{S}}P''D_{\Gamma_1})^{-1}=$$

$$=(zT_o-I-zD_{\Gamma_2^*}(\tilde{S}+zP''_*D_{\hat{S}^*}(z\hat{S}^*-I)^{-1}D_{\hat{S}}P'')D_{\Gamma_1}P)^{-1}=$$

$$=(zT_o-I+zD_{\Gamma_2^*}G(z)D_{\Gamma_1}P)^{-1},$$

and as in part (a) of the proof it follows that the g.r. of {T,S} generated by $\tilde{T}$, corresponds to the given function $G\in K(D_{\Gamma_1},D_{\Gamma_2^*})$.

The correspondence between the g.r. R_z of {T,S} and the functions $G\in K(D_{\Gamma_1},D_{\Gamma_2^*})$ is bijective (see part (b) of the proof). The last statement of Theorem 1 follows as in [8]. Indeed, given $G\in$ $\in K_o(D_{\Gamma_1},D_{\Gamma_2^*})$, we consider the operator $\tilde{T}:=T_o+D_{\Gamma_2^*}GD_{\Gamma_1}P$. According to [1] it is a canonical c.e. of {T,S} and, arguing as in (a), it follows that the corresponding g.r. is given by (9) with $G(z)\equiv G$. The theorem is proved.

REMARK. The relation (9) can also be written as

$$R_z=\overset{o}{R}_z(I+zD_{\Gamma_2^*}G(z)D_{\Gamma_1}P\overset{o}{R}_z)^{-1} \qquad (|z|<1)$$

(see, e.g., (10)).

4. In the description (9) of the g.r. of $\{T,S\}$ the canonical c.e. T_o plays a special role. It can be replaced by another canonical or even noncanonical c.e. Then the right hand side in (9) becomes slightly more complicated. In the following theorem we fix the c.e. $T_{\hat{G}}=:\hat{T}$ in $\hat{H}$, corresponding to the function $\hat{G}\in \mathcal{K}(\mathcal{D}_{\Gamma_1},\mathcal{D}_{\Gamma_2^*})$, and denote its g.r. by

$$\hat{R}_z:=P_H(z\hat{T}-\hat{I})^{-1}|_H \qquad (|z|<1).$$

THEOREM 2. *Let* $\{T,S\}$ *be a dual pair of contractions in the Hilbert space* H *and* $\hat{T}$ *an arbitrary c.e. of* $\{T,S\}$ *with corresponding function* $\hat{G}\in \mathcal{K}(\mathcal{D}_{\Gamma_1},\mathcal{D}_{\Gamma_2^*})$. *Then the formula*

$$\begin{aligned}R_z=\hat{R}_z-z\hat{R}_zD_{\Gamma_2^*}(G(z)-\hat{G}(z))(I-X_o(z)G(z))^{-1}\cdot\\ \cdot(I-X_o(z)\hat{G}(z))\cdot D_{\Gamma_1}P_{\mathcal{D}(T)^\perp}\hat{R}_z \qquad (|z|<1)\end{aligned} \tag{12}$$

establishes a 1,1-*correspondence between all g.r.* R_z *of* $\{T,S\}$ *and all functions* $G\in \mathcal{K}(\mathcal{D}_{\Gamma_1},\mathcal{D}_{\Gamma_2^*})$. *The g.r.* R_z *is canonical if and only if* $G\in \mathcal{K}_o(\mathcal{D}_{\Gamma_1},\mathcal{D}_{\Gamma_2^*})$.

PROOF. The Remark after Theorem 1 gives the representations

$$R_z=\mathring{R}_z(I+zD_{\Gamma_2^*}G(z)D_{\Gamma_1}P\mathring{R}_z)^{-1},$$

$$\hat{R}_z=\mathring{R}_z(I+zD_{\Gamma_2^*}\hat{G}(z)D_{\Gamma_1}P\mathring{R}_z)^{-1}, \qquad (|z|<1).$$

The second relation implies

$$\mathring{R}_z=\hat{R}_z(I-zD_{\Gamma_2^*}\hat{G}(z)D_{\Gamma_1}P\hat{R}_z)^{-1},$$

thus from the first relation we find

$$R_z=\hat{R}_z[I+zD_{\Gamma_2^*}(G(z)-\hat{G}(z))D_{\Gamma_1}P\hat{R}_z]^{-1},$$

which can easily seen to be equivalent to

$$R_z=\hat{R}_z-z\hat{R}_zD_{\Gamma_2^*}(G(z)-\hat{G}(z))[I+zD_{\Gamma_1}P\hat{R}_zD_{\Gamma_2^*}(G(z)-\hat{G}(z))]^{-1}D_{\Gamma_1}P\hat{R}_z . \tag{13}$$

If we put $\hat{Y}(z):=-zD_{\Gamma_1}P\hat{R}_zD_{\Gamma_2^*}$, it follows from (8) that the formula

$$\begin{aligned} X_o(z)&=-zD_{\Gamma_1}P(zT_o-I)^{-1}P_{\mathcal{D}(S)^\perp}D_{\Gamma_2^*}= \\ &=-zD_{\Gamma_1}P\hat{R}_z(I-zD_{\Gamma_2^*}\hat{G}(z)D_{\Gamma_1}P\hat{R}_z)^{-1}D_{\Gamma_2^*}= \\ &=-zD_{\Gamma_1}P\hat{R}_zD_{\Gamma_2^*}(I-z\hat{G}(z)D_{\Gamma_1}P\hat{R}_zD_{\Gamma_2^*})^{-1}= \\ &=\hat{Y}(z)(I+\hat{G}(z)\hat{Y}(z))^{-1} \end{aligned}$$

holds true, hence

$$\hat{Y}(z)=X_o(z)(I-\hat{G}(z)X_o(z))^{-1} . \tag{14}$$

Now (13) and (14) yield the relation

$$\begin{aligned} R_z&=\hat{R}_z-z\hat{R}_zD_{\Gamma_2^*}(G(z)-\hat{G}(z))\cdot \\ &\cdot[I-\hat{Y}(z)(G(z)-\hat{G}(z))]^{-1}D_{\Gamma_1}P\hat{R}_z= \\ &=\hat{R}_z-z\hat{R}_zD_{\Gamma_2^*}(G(z)-\hat{G}(z))\cdot \\ &\cdot[I-X_o(z)(I-\hat{G}(z)X_o(z))^{-1}(G(z)-\hat{G}(z))]^{-1}D_{\Gamma_1}P\hat{R}_z , \end{aligned} \tag{15}$$

which coincides with (12). The statement now follows immediately from Theorem 1.

The first equality in (15) gives an alternative description of the g.r. of the dual pair {T,S}, using the function $\hat{Y}$ (which is connected with the c.e. $\hat{T}$) instead of X_o. In general, however, $\hat{Y}$ is not a contractive function. If we choose a *canonical* c.e. $\hat{T}$ with the additional property that for the corresponding contraction $\hat{G}\in[\mathcal{D}_{\Gamma_1},\mathcal{D}_{\Gamma_2^*}]$ the operators $D_{\hat{G}},D_{\hat{G}^*}$ are invertible, then a description of the g.r. can be given using the contractive analytic function $\hat{X}$:

$$\text{(16)}\qquad \begin{aligned}\hat{X}(z):=X_{\hat{G}}(z)=-\hat{G}^*-zD_{\hat{G}}D_{\Gamma_1}P(z\hat{T}-I)^{-1}D_{\Gamma_2^*}D_{\hat{G}^*}=\\ =\hat{G}^*+D_{\hat{G}}\hat{Y}(z)D_{\hat{G}^*}\qquad (|z|<1);\end{aligned}$$

see Corollary 2. We give this description only in case that $\hat{G}$ is a contraction with $||\hat{G}||<1$.

Let $\hat{T}$ *in Theorem* 2 *be chosen such that for the corresponding contraction* $\hat{G}\varepsilon[\mathcal{D}_{\Gamma_1},\mathcal{D}_{\Gamma_2^*}]$ *we have* $||\hat{G}||<1$. *Then the formula*

$$R_z=\hat{R}_z-z\hat{R}_zD_{\Gamma_2^*}D_{\hat{G}^*}H(z)(I-\hat{X}(z)H(z))^{-1}D_{\hat{G}}D_{\Gamma_1}P_{\mathcal{D}(T)^\perp}\hat{R}_z \qquad (|z|<1)$$

establishes a 1,1-*correspondence between all g.r.* R_z *of* $\{T,S\}$ *and all functions* $H\varepsilon K(\mathcal{D}_{\Gamma_1},\mathcal{D}_{\Gamma_2^*})$. *The g.r.* R_z *is canonical if and only if* $H\varepsilon K_o(\mathcal{D}_{\Gamma_1},\mathcal{D}_{\Gamma_2^*})$.

Here, of course, we can write $\mathcal{D}_{\hat{G}}$ and $\mathcal{D}_{\hat{G}^*}$ instead of $\mathcal{D}_{\Gamma_1}$ and $\mathcal{D}_{\Gamma_2^*}$ respectively.

This statement follows from Theorem 2, if it is shown that the relation

$$\text{(17)}\qquad \begin{aligned}(G(z)-\hat{G})(I-X_o(z)G(z))^{-1}(I-X_o(z)\hat{G})=\\ =D_{\hat{G}^*}H(z)(I-\hat{X}(z)H(z))^{-1}D_{\hat{G}}\end{aligned}$$

holds with

$$H(z):=D_{\hat{G}^*}^{-1}(G(z)-\hat{G})(I-\hat{G}^*G(z))^{-1}D_{\hat{G}}\ ,$$

and that $H(z)$ is a contraction ($|z|<1$). The formula (17) can be checked using (14) and (16). Moreover, the inequality $G(z)^*G(z)\le$ $\le I$ gives

$$(G(z)^*-\hat{G}^*)D_{\hat{G}^*}^{-2}(G(z)-\hat{G})\le(I-G(z)^*\hat{G})D_{\hat{G}}^{-2}(I-\hat{G}^*G(z))\ ,$$

which is equivalent to $H(z)^*H(z)\le I$.

5. The Theorem in [8] is a special case of Theorem 1. Indeed if a nondensely defined contraction T in H is given, we can consider the dual pair $\{T,S_o\}$ with $S_o=0$ on $\mathcal{D}(S_o)=0$. It follows

$$A:\mathcal{D}(T)\to\{0\},\qquad D_A=I|_{\mathcal{D}(T)},$$

$$\Gamma_1:\mathcal{D}(T)^{\perp}\to\{0\},\qquad D_{\Gamma_1}=I|_{\mathcal{D}(T)^{\perp}},$$

$$\Gamma_2=\overset{o}{T},\qquad D_{\Gamma_2^*}=D_{\overset{o}{T}^*},\qquad (\overset{o}{T}\supset T,\ \overset{o}{T}|_{\mathcal{D}(T)^{\perp}}=0),$$

and Theorem 1 turns into the Theorem in [8]. Moreover, Theorem 2 implies the following description of the g.r. of T if an arbitrary c.e. $\hat{T}$ of T is fixed[1]. Here $\hat{G}\in K(\mathcal{D}(T)^{\perp},\mathcal{D}_{\overset{o}{T}^*})$ denotes again the "parameter" corresponding to the g.r. $\hat{R}_z=P_H(z\hat{T}-\hat{I})^{-1}|_H$ according to Theorem 1.

THEOREM 3. *Let* T *be a nondensely defined contraction in the Hilbert space* H *and* $\hat{T}$ *an arbitrary c.e. of* T. *Then the formula*

$$R_z=\hat{R}_z-z\hat{R}_zD_{\overset{o}{T}^*}(G(z)-\hat{G}(z))(I-X_o(z)G(z))^{-1}\cdot \tag{18}$$
$$\cdot(I-X_o(z)\hat{G}(z))P_{\mathcal{D}(T)^{\perp}}\hat{R}_z\qquad(|z|<1)$$

establishes a 1,1-*correspondence between all g.r.* R_z *of* T *and all functions* $G\in K(\mathcal{D}(T)^{\perp},\mathcal{D}_{\overset{o}{T}^*})$. *The g.r.* R_z *is canonical if and only if* $G\in K_o(\mathcal{D}(T)^{\perp},\mathcal{D}_{\overset{o}{T}^*})$.

In the following we consider the particular case that T is an isometric operator in H. Then $D_{\overset{o}{T}^*}$ is the orthogonal projection onto $R(T)^{\perp}$. If $\tilde{T}$ is an arbitrary c.e. of T, then any unitary dilation U of $\tilde{T}$ is also a c.e. of T. Indeed, we have for $x\in\mathcal{D}(T)$:

$$||Ux-Tx||^2=(Ux,Ux)-2\mathrm{Re}(Ux,Tx)+||Tx||^2=$$
$$=||x||^2-2\mathrm{Re}(\tilde{T}x,Tx)+||x||^2=0.$$

Now we choose the c.e. $\hat{T}$ of T to be a unitary operator in $\hat{H}\supset H$.

[1] A c.e. (g.r.) of $\{T,S_o\}$ is called a c.e. (g.r. respectively) of T.

Then the operators $\hat{R}_z D_{\overset{o}{T}*}$ and $P_{\mathcal{D}(T)^\perp}\hat{R}_z$, which appear in (18), coincide with the mappings $-\Gamma_{1/z,a}$ and $-\Gamma^*_{\bar{z},i}$ introduced in [9, p.9]. Thus the description of the g.r. in Theorem 3 gives Satz 2.6 of [9] for the special case of a Hilbert space ($\varkappa=0$). Here the g.r. R_z of T, generated by a c.e. $\tilde{T}$ of T, coincides (for $|z|<1$) with the g.r. of T generated by an arbitrary unitary dilation of $\tilde{T}$.

Finally we shall explain the connection of Theorem 3 with Satz 4.1 in [5]. To this end let T be an isometric operator in $\mathcal{H}$ with equal (and non-zero) defect numbers. We choose a canonical unitary extension $U=\hat{T}$. It admits a representation

$$U=\overset{o}{T}+P_{\mathcal{R}(T)^\perp}G_o P_{\mathcal{D}(T)^\perp}$$

with some unitary operator G_o from $\mathcal{D}(T)^\perp$ onto $\mathcal{R}(T)^\perp$. The relation

$$G(z)=G_o E(z) \qquad (|z|<1)$$

establishes a 1,1-correspondence between all $G\epsilon K(\mathcal{D}(T)^\perp, \mathcal{R}(T)^\perp)$ and all $E\epsilon K(\mathcal{D}(T)^\perp,\mathcal{D}(T)^\perp)$. Moreover, for the operators Γ_z , introduced in [5, (4.1)] we find

$$\Gamma_{\frac{1}{\bar{z}}}=z\hat{R}_z G_o \ , \quad \Gamma^*_{\bar{z}}=-P_{\mathcal{D}(T)^\perp}\hat{R}_z \ ,$$

and for the "characteristic function" X(z) of [5, p.386]

$$X(\bar{z})^*=-zP_{\mathcal{D}(T)^\perp}\overset{o}{R}_z G_o=X_o(z)G_o \ .$$

Now it is easy to see that in this special case (18) coincides with [5, (4.4)], that is, for the case of a Hilbert space the Satz 4.1 in [5] is an immediate consequence of Theorem 3.

REFERENCES

1. Arsene, Gr.; Gheondea, A.: Completing matrix contractions, *J.Operator Theory* 7 (1982), 179-189.

2. Bognár, J.: *Indefinite inner product spaces*, Springer-Verlag, 1974.

3. Crandall, M.G.: Norm preserving extensions of linear

transformations on Hilbert spaces, *Proc.Amer.Math.Soc.* 21 (1969), 335-340.

4. Kreĭn, M.G.: The theory of self-adjoint extensions of semi-bounded Hermitian transformations and its applications, (Russian), *Mat.Sb.* 20 (1947), 431-495; 21 (1947), 365-404.

5. Kreĭn, M.G.; Langer, H.: Über die verallgemeinerten Resolventen und die charakteristische Funktion eines isometrischen Operators im Raume $\Pi_\varkappa$, *Colloquia mathematica societatis János Bolyai*, 5: Hilbert space operators and operator algebras (Tihany 1970), North-Holland Publishing Company, 1972, pp.353-399.

6. Kreĭn, M.G.; Langer, H.: Über einige Fortsetzungsprobleme, die eng mit der Theorie hermitescher Operatoren im Raume $\Pi_\varkappa$ zusammenhängen. I. Einige Funktionenklassen und ihre Darstellungen, *Math.Nachr.* 77(1977), 187-236.

7. Kreĭn, M.G.; Ovčarenko, I.Je.: On the Q-functions and the sc-resolvents of non-densely defined Hermitian contractions (Russian), *Sibirsk.Math.Ž.* 18(1977), 1032-1056.

8. Langer, H.; Textorius, B.: Generalized resolvents of contractions, *Acta Sci.Math. (Szeged)*, to appear.

9. Langer, H.; Sorjonen, P.: Verallgemeinerte Resolventen hermitescher und isometrischer Operatoren im Pontrjaginraum, *Ann.Acad.Sci.Fenn.* A I, 561 (1974), 3-45.

10. Parrott, S.: On a quotient norm and the Sz.-Nagy-Foiaş lifting theorem, *J.Functional Analysis* 30 (1978), 311-328.

11. Phillips, R.S.: The extension of dual subspaces invariant under an algebra, *Proc.Internat.Symp.Linear Algebra, Israel*, 1960, Academic Press, 1961, pp.366-398.

12. Sz.-Nagy, B.; Foiaş, C.: *Harmonic analysis of operators on Hilbert space*, North Holland Publishing Company, 1970.

H. Langer
Technical University,
Sektion Mathematik,
DDR-8027, Dresden,
G.D.R.

B.Textorius
Linköping University,
Department of Mathematics,
S-58183 Linköping,
Sweden.

ANNIHILATORS OF OPERATOR ALGEBRAS

David R.Larson

The structure of the predual of an ultraweakly closed operator algebra can be very revealing of internal structural properties of the algebra. This relationship has been most important in the theory of von Neumann algebras, and has recently become significant in the study of more general ultraweakly closed algebras.

The purpose of this note is to show that, analogously, knowledge of the *annihilator* of such an algebra can be revealing of *external* structural properties of the algebra and seems to lend perspective to certain open questions. In particular, the notions of reflexivity, n-reflexivity, and the Arveson distance estimate have natural formulations in this setting.

Let H be separable Hilbert space. The usual notation Lat(A) will denote the lattice of invariant subspaces (or projections) for a subset $A \subseteq L(H)$, and Alg(L) will denote the algebra of bounded linear operators leaving invariant every member of a family L of subspaces (or projections). A is *reflexive* if $A=\mathrm{AlgLat}(A)$ and L is reflexive if $L=\mathrm{LatAlg}(L)$. An operator T is said to be reflexive if the weakly closed algebra generated by T and I, denoted W(T), is reflexive. T is said to be *n-reflexive* if the n-fold inflation $T^{(n)}=T \oplus T \oplus \ldots \oplus T$, acting on $H^{(n)}=H \oplus \ldots \oplus H$, is reflexive, and an algebra A is n-reflexive if $A^{(n)}=\{T^{(n)}:T\varepsilon A\}$ is reflexive. If $\{A_\lambda\}$ is a family of algebras then $\bigcap_\lambda \mathrm{Lat}(A_\lambda)=\mathrm{Lat}(\bigcup_\lambda A_\lambda)$, and if $\{L_\lambda\}$ is a family of lattices then $\bigcap_\lambda \mathrm{Alg}(L_\lambda)=\mathrm{Alg}(\bigcup_\lambda L_\lambda)$. So arbitrary intersections of reflexive algebras (lattices) are reflexive. The same is not true of joins.

Let L_* denote the trace class operators in L(H). Then L(H) is identified with $(L_*)^*$ via the pairing $(f,T)=\mathrm{Tr}(Tf)$, $f\varepsilon L_*$,

$T \in L(H)$. Let K denote the compact operators in $L(H)$. Then the same pairing identifies L_* with $(K)^*$. Let $||f||_1$ denote the trace class norm of f in L_*. By *annih*(A) we mean the set of all trace-class operators f such that $Tr(Af)=0$, $A \in A$. The ultraweak (σ-weak) topology on $L(H)$ is the w^*-topology under the identification $L(H)= =(L_*)^*$. If A is σ-weakly closed then A is identified as the dual of the quotient $L_*/\text{annih } A$. We write $\text{predual}(A)=L_*/\text{annih}(A)$. If $\{S_\lambda : \lambda \in \Lambda\}$ is a family of σ-weakly closed linear subspaces of $L(H)$ then $\text{annih}(\bigcap_\lambda S_\lambda)=\overline{\text{span}}\{\bigcup_\lambda \text{annih}(S_\lambda)\}$, where closure denotes $||\cdot||_1$ closure, and $\text{annih}(\bigcup_\lambda S_\lambda)=\bigcap_\lambda \text{annih}(S_\lambda)$.

If $x,y \in H$ we denote the rank-1 operator $u \to (u,x)y$ by $x \otimes y$. If A is an algebra containing I and $P \in \text{Lat}(A)$ with $P \neq 0$, I let x, y be nonzero vectors in $P^\perp H$, PH, respectively. If $f=x \otimes y$ then $f \in \text{annih}(A)$ since $Tr(Af)=(Ay,x)=0$, $A \in A$. Conversely, if $f=x \otimes y$ is a rank-1 operator in annih(A) then $(Ay,x)=0$, $A \in A$, so $[Ay]$ is a nontrivial invariant subspace for A. Thus A has a nontrivial invariant subspace iff annih(A) contains a rank-1 operator. More can be said in this direction.

LEMMA 1. *If* A *is an arbitrary operator algebra containing* I *then* annih(A) *and* annih(AlgLat A) *contain the same* rank-1 *operators. If* f *is* rank-1, *then* $f \in \text{annih}(A)$ *if and only if* $f=PfP^\perp$ *for some* $P \in \text{Lat}(A)$.

PROOF. Suppose $f=x \otimes y \in \text{annih}(A)$. Then for $A \in A$ we have $0= =Tr(Af)=(Ay,x)$. Let P be the projection onto the subspace $[Ay]$. If $B \in \text{AlgLat } A$ then $BPH \subseteq PH$, and $P \perp x$, so $Tr(Bf)=(By,x)=0$. So $f \in \in \text{annih}(\text{AlgLat } A)$.

For the last sentence, if f is a trace class operator with $f=PfP^\perp$ for $P \in \text{Lat}(A)$ then $f \in \text{annih}(A)$ trivially. The converse follows from above. ■

LEMMA 2. *Let* A *be a* σ*-weakly closed algebra containing* I. *Then* A *is reflexive if and only if* annih(A) *is the* $||\cdot||_1$*-closed linear span of* rank-1 *operators.*

PROOF. We have $\text{AlgLat}(A) \supseteq A$ so $\text{annih}(\text{AlgLat } A) \subseteq \text{annih}(A)$. If annih(A) is generated by rank-1 operators this last inclusion is an equality by Lemma 1, so AlgLat $A=A$.

Conversely, suppose A is reflexive and let $L=\mathrm{Lat}(A)$. For each $P\varepsilon L$ let $L_p=\{0,P,I\}$. Then $A=\cap\{\mathrm{Alg}(L_p):P\varepsilon L\}$. It is easily verified that $\mathrm{annih}(\mathrm{Alg}(L_p))=PL_*P^\perp$ which is the closed linear span of its rank-1 elements since L_* is. Since annih(A) is the $||\cdot||_1$-closed span of $\cup\{\mathrm{annih}(\mathrm{Alg}L_p):P\varepsilon L\}$ it also is generated by rank-1 elements. ■

While n-reflexivity for n>1 does not imply the existence of a single nontrivial invariant subspace, it remains an interesting structural property that has a natural characterization in terms of the annihilator.

LEMMA 3. *Let* A *be a* σ*-weakly closed linear subspace of* L(H), *and let* $1\le n<\infty$. *Then* $\mathrm{annih}(A^{(n)})$ *is the set of all trace class operators* $f=(f_{ij})_{i,j=1}^n$ *in* $L(H^{(n)})$ *such that* $\sum_{i=1}^n f_{ii}\varepsilon\,\mathrm{annih}(A)$.

PROOF. For $A\varepsilon L(H)$ we have $\mathrm{Tr}(A^{(n)}f)=\sum_{i=1}^n \mathrm{Tr}(Af_{ii})$. The conclu- clusion follows.■

THEOREM 4. *Let* A *be a* σ*-weakly closed subalgebra of* L(H) *containing* I *and let* $1\le n<\infty$. *Then* A *is* n*-reflexive if and only if* annih(A) *is the trace class norm closed linear span of operators of* rank≤n.

PROOF. Let S be the trace-class norm closed linear span of operators in $\mathrm{annih}(A^{(n)})$ of rank≤1. We will show that $S=\mathrm{annih}(A^{(n)})$, thus proving reflexivity. Let $f=(f_{ij})$ be an element of $\mathrm{annih}(A^{(n)})$, so $\Sigma f_{ii}\varepsilon\,\mathrm{annih}(A)$. Every zero-diagonal operator matrix with trace class entries is clearly in S, thus we reduce to the case in which f is diagonal. We use the notation $f=\mathrm{diag}(f_{11},f_{22},\ldots,f_{nn})$.

First suppose that each f_{ii} has rank≤1. Write $f_{ii}=x_i\otimes y_i$ for vectors $x_i,y_i\varepsilon H$. Let $g_{ij}=x_j\otimes y_i$, $1\le i,\ j\le n$ and let $g=(g_{ij})$. Then $g\varepsilon\,\mathrm{annih}(A^{(n)})$, and the range of g is the span of the vector $y_1\oplus y_2\oplus\ldots\oplus y_n$ so g has rank≤1. By a standard argument f is a convex combination of operators unitarily equivalent to g having the same diagonal. These operators are in $\mathrm{annih}(A^{(n)})$, which proves $f\varepsilon S$.

For f_{ii} of arbitrary rank proceed as follows. For $1\le i\le n-1$

let h_i be the diagonal matrix with i,i-element equal to $f_{11}+f_{22}+\ldots$ $\ldots+f_{ii}$, i+1, i+1-element equal to $-(f_{11}+f_{22}+\ldots+f_{ii})$, and all other elements 0. Let h be the diagonal matrix with n,n-element $f_{11}+\ldots+f_{nn}$ and all other elements 0. Then h_i, $h\varepsilon\,\text{annih}(A^{(n)})$, and $f=h_1+\ldots+h_{n-1}+h$. From Lemma 3 it is clear that $h_i\varepsilon S$, $1\le i\le n-1$. We must show $h\varepsilon S$.

Since $f_{11}+\ldots+f_{nn}\varepsilon\,\text{annih}(A)$, by hypothesis this can be approximated by finite sums of operators in annih(A) of rank≤n. Thus it will suffice to show that every operator of the form diag(0,0,. ...,0,r) with $r\varepsilon\,\text{annih}(A)$, rank(r)≤n, is in S. Write $r=r_1+\ldots+r_n$ where $\text{rank}(r_i)\le 1$, r_i not necessarily in annih(A). Now write $\tilde f_{ii}=$ $=r_i$ and $\tilde f=\text{diag}(\tilde f_{11},\tilde f_{22},\ldots,\tilde f_{nn})$. Then $\tilde f\varepsilon\,\text{annih}(A^{(n)})$, and $\tilde f\varepsilon S$ since $\text{rank}(\tilde f_{ii})\le 1$. Now repeat the argument in the previous paragraph replacing f with $\tilde f$ and f_{ii} with $\tilde f_{ii}$, obtaining $\tilde f=\tilde h_1+\tilde h_2+\ldots+\tilde h_{n-1}+$ $+\tilde h$ with $\tilde f$, $\tilde h_i\varepsilon S$ and $h=\text{diag}(0,0,\ldots,r)$. Hence $h\varepsilon S$. Thus $A^{(n)}$ is reflexive.

For the converse, assume $A^{(n)}$ is reflexive. Let $r\varepsilon\,\text{annih}(A)$ and let $R=\text{diag}(0,0,\ldots,0,r)$. We have $R=\lim_{\ell} R_\ell$ where R_ℓ is a finite sum of operator in $\text{annih}(A^{(n)})$ of rank≤1.

Write $R_\ell=\sum_k R^k_\ell$. Now let $R^k_{\ell ij}$ denote the i,j-matrix element of R^k_ℓ, and let $R_{\ell ij}$ denote the i,j-matrix element of R_ℓ. We have $\lim_\ell R_{\ell nn}=r$ and $\lim_\ell R_{\ell ii}=0$, $1\le i\le n-1$, so $\lim_\ell \sum_i R_{\ell ii}=r$. Also, $R_{\ell ii}=$ $=\sum_k R^k_{\ell ii}$, so $r=\lim_\ell\sum_k\sum_i R^k_{\ell ii}$. Each sum $\sum_{i=1}^n R^k_{\ell ii}$ is in annih(A) and has rank≤n.■

COROLLARY 5. *If an algebra* A *is* n-*reflexive then it is* m-*reflexive for all* m≥n.

Let A be σ-weakly closed operator algebra containing I. For $1\le n<\infty$ note that $\text{AlgLat}(A^{(n)})$ is contained in $(L(H))^{(n)}$ since the latter is reflexive and let $A_n=\{A\varepsilon L(H):A^{(n)}\varepsilon\,\text{AlgLat}(A^{(n)})\}$. Then A_n is the smallest n-reflexive algebra containing A. We say that A_n is the n-reflexive algebra generated by A. If A is an algebra containing I and $1\le n<\infty$ let E_n be the $||\cdot||_1$-closed span of the operators in annih(A) of rank$\le$n. While it may not seem obvious that the polar of E_n in $L(H)$ is an algebra, this is indeed the case and we have $A_n=(E_n)^{\perp}$.

COROLLARY 6. *The annihilator of* A_n *is the trace-class norm closed linear span of the operators in* annih(A) *of rank*$\le$n.

PROOF. By (2), (3) and (4), annih(A_n) is the $||\cdot||_1$-closed linear span of operators $\sum_{i=1}^{n} f_{ii}$ where the operator matrix (f_{ij}) is a rank-1 operator in the annihilator of AlgLat($A^{(n)}$). Each Σf_{ii} has rank$\le$n and is contained in annih(A) since $A\subseteq A_n$. Conversely, if $g\in$annih(A) has rank$\le$n write $g=g_1+\ldots+g_n$ where rank$(g_i)\le 1$, g_i not necessarily in annih(A). Then $G=\text{diag}(g_1,g_2,\ldots,g_n)\in\text{annih}(A^{(n)})$. As in the proof of Theorem 4 there exists a rank-1 operator with the same diagonal as G hence in annih($A^{(n)}$). By (2) this is also in annih(AlgLat($A^{(n)}$)). Thus $g=\Sigma g_i\in\text{annih}(A_n)$.■

REMARK. It is easily proven from duality considerations that a σ-weakly closed linear subspace of $L(H)$ is weakly closed if and only if its annihilator is the $||\cdot||_1$-closed linear span of finite rank operators. If A is a weakly closed algebra with I and $T\in L(H)$ it is well known that $\text{dist}(T,A)=\lim_n \text{dist}(T^{(n)},\text{AlgLat}A^{(n)})=\lim_n \text{dist}(T,A_n)$. This can be seen intuitively from Corollary 6 since annih(A) is the closed span of finite rank elements and the distance from T to any σ-weakly closed algebra B is $\sup\{|\text{Tr}(Tf)|:f\in\text{annih}B,||f||_1\le$ $\le 1\}$. For any algebra A containing I the weak closure of A is $\bigcap_n A_n$. These annihilator considerations seem to point out the special role of the weakly closed algebras in the general theory of ultraweakly closed algebras. In particular it can become of interest to know when, for a given operator T, the σ-weakly closed algebra generated by T and I is in fact weakly closed. Questions of this nature do not seem to have been pursued in the literature. The following simple application of Theorem 4 suggests that these questions could be interesting.

COROLLARY 7. *Let* A *be a weakly closed algebra containing* I. *Suppose every element of* L_*/annih(A) *has the form* f+annih(A) *with* rank(f)$\le$1. *Then* A *is* 3-*reflexive*.

PROOF. A is weakly closed so annih(A) is the closed span of finite rank elements. Let g be a finite rank element of annih(A) and write $g=g_1+\ldots+g_m$ where g_i is rank-1, not necessarily in annih(A). By hypothesis there exist $h_1,\ldots,h_{m-1}$ with rank$(h_i)\leq 1$ such that each term in the decomposition $g=(g_1+g_2-h_1)+(h_1+g_3-h_2)+ +\ldots+(h_{m-2}+g_m-h_{m-1})+h_{m-1}$ is in annih(A). Then A is 3-reflexive by Theorem 4.■

Let us say that an algebra $A\subset L(H)$ has property P_1 if it has the predual property in Corollary 6: every element of $L_*/\text{annih}(A)$ has the form $f+\text{annih}(A)$ with rank$(f)\leq 1$. Corollary 7 says that a *weakly* closed algebra containing I with property P_1 is 3-reflexive. It was shown by Brown, Chevreau and Pearcy in [2] that completely nonunitary contractions with "rich spectrum" not only have nontrivial invariant subspaces, but they showed that such an operator T which could not be eliminated as having an invariant subspace by preliminary means had the property that the σ-weakly closed algebra generated by $\{T,I\}$ has the predual property P_1.

QUESTION 1. If T is a completely nonunitary contraction with "rich spectrum" is the σ-weakly closed algebra generated by $\{T,I\}$ closed? If so T would be 3-reflexive. This would be interesting structurally, although we note that this property above would not to our knowledge directly imply the existence of an invariant subspace. Rich spectrum yielded much stronger properties that were used in [2] to conclude that T was not transitive.

A strengthening of Q1 would be:

QUESTION 2. Is a σ-weakly closed algebra with property P_1 necessarily weakly closed?

QUESTION 3. Does a weakly closed algebra with property P_1 (hence 3-reflexive) necessarily have a nontrivial invariant subspace? Equivalently, does annih(A) contain rank-1 operators?

REMARK. Corollary 7 has a natural generalization which we present for completeness. Let us say that an algebra A has property P_n if every element of $L_*/\text{annih}(A)$ has the form $f+\text{annih}(A)$ with rank$(f)\leq n$.

COROLLARY 8. *Let* A *be a weakly closed algebra containing* I *which has property* P_n. *Then* A *is* 3n-*reflexive.*

REMARK. It is in actuality somewhat missleading to refer to P_n as a predual property since it depends on the representation of a predual of a σ-weakly closed algebra as $L_*/\text{annih}(A)$ and is not independent of the particular representation of A as a σ-weakly closed algebra on Hilbert space.

QUESTION 4. Let A be a σ-weakly closed algebra containing I and let F_1 be the set of elements of $L_*/\text{annih}(A)$ which have the form $f+\text{annih}(A)$ with $\text{rank}(f)\leq 1$. Is F_1 necessarily closed in the quotient norm of $L_*/\text{annih}(A)$? We feel that the answer is likely negative without strong conditions imposed on A. Can one find reasonable general conditions? The motivation for this question comes from [2]. If it were known independently that F_1 is closed if A is the σ-weakly closed algebra generated by $\{T,I\}$ and T is a completely nonunitary contraction with rich spectrum, then the proof in [2] could have been simplified. The beautiful argument in [2] shows that F_1 is closed by in fact proving that it is all of $L_*/\text{annih } A$.

REMARK. It is known that a weakly closed algebra containing I with property P_1 need not be reflexive, even in finite dimensions. However we do have the following:

LEMMA 9. *Let* A *be a reflexive algebra with property* P_1. *Then every* σ-*weakly closed subalgebra containing* I *is also reflexive.*

PROOF. Let B be a σ-weakly closed subalgebra of A, and let $\tilde{B}=\text{AlgLat } B$. Then $\tilde{B}\subsetneq A$ since A is reflexive. If $B\neq\tilde{B}$ there exists an element of $L*/\text{annih}(A)$ that annihilates B but does not annihilate $\tilde{B}$. This has the form $f+\text{annih}(A)$ with $\text{rank}(f)=1$. But then $f\in\text{annih}(B)$ so by Lemma 1 $f\in\text{annih}(\tilde{B})$, a contradiction.∎

REMARK. The above proof easily extends to show that if A is an n-reflexive algebra with property P_n then every σ-weakly closed subalgebra containing I is n-reflexive.

The annihilator of an algebra which is a commutant has a very simple characterization.

LEMMA 10. *Let* $\mathcal{A}$ *be a subset of* L(H). *Then the annihilator of the commutant* $\mathcal{A}'$ *is the trace-class norm closed span of* $\{Af-fA: A\in\mathcal{A},\ f\in\mathcal{L}_*\}$.

PROOF. If $A\in\mathcal{A}$, $f\in\mathcal{L}_*$, let $g=Af-fA$. Then for $B\in\mathcal{A}'$, $Tr(Bg)=Tr((BA-AB)f)=0$ so $g\in annih(\mathcal{A}')$. Conversely, let B be an operator for which $Tr(B(Af-fA))=0$, $A\in\mathcal{A}$, $f\in\mathcal{L}_*$. Then $Tr((BA-AB)f)=0$, $f\in\mathcal{L}_*$, so $BA-AB=0$.■

COROLLARY 11. *Every commutant is a* 2-*reflexive algebra.*

PROOF. The above lemma shows that the annihilator of a commutant is generated by rank$\le$2 operators.■

REMARK. The above shows that if there exists a transitive operator then there exists a nontrivial 2-reflexive transitive algebra: namely its commutant. More generally, if there exists a transitive operator algebra with nontrivial commutant then there exists a nontrivial 2-reflexive transitive algebra: its double commutant.

QUESTION 5. Can it be shown that the existence of any nontrivial weakly closed transitive algebra would imply the existence of a nontrivial 2-reflexive transitive algebra? Note that it would imply the existence of a nontrivial n-reflexive transitive algebra for some finite n. Would the existence of a transitive operator imply the existence of a 2-reflexive transitive operator?

REMARK. Corollary 11 shows that while reflexive operators and algebras are rather special the larger class of 2-reflexive operators and algebras is significantly richer. Let W(A) denote the weakly closed algebra generated by $\{A,I\}$. If $W(A)=\{A\}''$ then A is 2-reflexive. In particular, the Volterra operator is 2-reflexive, although it is unicellular so it is far from being reflexive.

QUESTION 6. Which weighted shifts are 2-reflexive? Is there a reasonable classification of compact 2-reflexive operators?

REMARK. If T is an operator the set $\{Tf-fT: f\in\mathcal{L}_*\}$ is a linear space, so the question of whether T has a nontrivial hyperinvariant subspace translates to the question of whether the $||\cdot||_1$-closure of this space contains a rank-1 operator.

The present investigation was motivated in part by work of W.Arveson [1] and by independent work of C.Lance [7] in which a formula for the distance from an arbitrary operator to a given nest algebra is obtained in terms of projections in the nest. This was extended by several authors [3,4,5,6]. The basic question remains: for which reflexive algebras A (if not for all) is there a constant K such that for every $T\varepsilon L(H)$ we have $\operatorname{dist}(T,A)\le K\cdot\sup\{||P^{\perp}TP||:P\varepsilon \operatorname{Lat} A\}$? We show that this question has a natural formulation in terms of the annihilator of A which seems to illuminate the question and raises some related questions. After obtaining this it became known to us that this formulation had been obtained by E.Christensen [4] for the case in which A is a von Neumann algebra.

PROPOSITION 12. *Let* A *be a reflexive algebra. Let* $B_1=\{f\varepsilon \operatorname{annih} A: ||f||_1\le 1\}$, *and let* C_1 *denote the* $||\cdot||_1$*-closed convex hull of the operators in* B_1 *of rank*≤ 1. *Then there exists a constant* K *such that for all* $T\varepsilon L(H)$ *we have* $\operatorname{dist}(T,A)\le K\cdot\sup\{||P^{\perp}TP||:P\varepsilon \operatorname{Lat} A\}$ *if and only if* C_1 *has nonempty relative interior in* annih(A). *In this case the smallest constant* K *for which this estimate holds is* $1/R$ *where* R *is the largest radius such that* $\{f\varepsilon \operatorname{annih} A: ||f||_1\le R\}\subseteq C_1$.

PROOF. We may assume $A\ne L(H)$. Suppose C_1 has nonempty relative interior E. Then since C_1 is closed and balanced it contains a ball of maximal radius $R>0$ with center 0 in annih(A). Let $k=1/R$. Let $T\varepsilon L(H)$ and let $d=\operatorname{dist}(T,A)$. We have $d=\sup\{|\operatorname{Tr}(Tf)|: f\varepsilon B_1\}$ from the duality, and since $kC_1\supseteq B_1$ we have

$$d\le k\sup\{|\operatorname{Tr}(Tf)|: f\varepsilon C_1\}=k\sup\{|\operatorname{Tr}(Tf)|: f\varepsilon B_1 , f \text{ of rank-1}\}.$$

So by Lemma 1 we have

$$d\le k\sup\{|\operatorname{Tr}(TPfP^{\perp})|: f \text{ of rank-1}, ||f||\le 1, P\varepsilon \operatorname{Lat} A\}=$$
$$=k\sup\{|\operatorname{Tr}(P^{\perp}TPf)|: f \text{ of rank-1}, ||f||\le 1, P\varepsilon \operatorname{Lat} A\}=$$
$$=k\sup\{||P^{\perp}TP||: P\varepsilon \operatorname{Lat}(A)\},$$

as desired.

Conversely, suppose a distance estimate holds. The set of constants for which an estimate holds is closed so contains a minimal element K. Let $r=1/K$. Suppose there exists $f\varepsilon \text{annih}(A)$ with $||f||_1 \le r$ and $f \notin C_1$. By a standard separation theorem there exists an element ϕ in the dual of annih(A) of norm 1 such that $\text{Re}\phi(g) < \text{Re}\phi(f)$, $g\varepsilon C_1$. Since $0\varepsilon C_1$ we must have $\text{Re}\phi(f)>0$. Since C_1 is invariant under multiplication by scalars of modulus 1 we have that

$$|\phi(g)| < \text{Re}\phi(f) \le |\phi(f)|, \qquad g\varepsilon C_1 .$$

Since annih(A) is a closed subspace of the Banach space L_* its dual is identified isometrically with $L(H)/A$. If $T\varepsilon L(H)$ the norm of its image in this quotient is $\text{dist}(T,A)$. Thus there exists $T\varepsilon L(H)$ with $\text{dist}(T,A)=1$ such that

$$|\text{Tr}(Tg)| < |\text{Tr}(Tf)|, \qquad g\varepsilon C_1 .$$

Choose $P_n \varepsilon L=\text{Lat}(A)$ such that $\lim ||P_n^\perp T P_n|| \ge r\,\text{dist}(T,A)=r$, and choose unit vectors $u_n \varepsilon P_n^\perp H$, $v_n \varepsilon P_n H$, such that $(Tv_n,u_n)=||P_n^\perp T P_n||$. Let $g_n=u_n \otimes v_n$. Then $g_n \varepsilon \text{annih}(A)$, $||g_n||_1=1$, so $g_n \varepsilon C_1$, and $|\text{Tr}(Tg_n)|=|(Tv_n,u_n)|=||P_n^\perp T P_n||$. Thus $\sup\{|\text{Tr}(Tg)|: g\varepsilon C_1\} \ge r$. Hence $|\text{Tr}(Tf)|>r$, a contradiction since $f\varepsilon \text{annih}(A)$, $||f||_1 \le r$, and $\text{dist}(T,A)=1$. So $C_1 \supseteq \{f\varepsilon \text{annih}(A): ||f||_1 \le r\}$. Thus C_1 has nonempty relative interior. The above shows that $K \le 1/R$ and also $R \ge 1/K$; hence $K=1/R$. ■

REMARKS. The ideas in the proof of Proposition 12 were implicit in the proofs in [1] and especially in [7] for nest algebras. These showed essentially that every norm 1 trace class operator in the annihilator of a nest algebra could be approximated in norm by convex combinations of rank-1 operators of norm ≤ 1 from the annihilator.

In finite dimensions every reflexive algebra satisfies an Arveson distance estimate. The constant need not be 1. In infinite dimensions, it is known [4,6] that a von Neumann algebra satisfies an Arveson distance estimate if and only if every derivation from it into L(H) is inner. As pointed out in [5], if every member of a class of reflexive algebras closed under countable direct sums satisfies an Arveson distance estimate then there must be a universal constant for the class. No example is known of a reflexive algebra which fails to satisfy a distance estimate. However, the class of algebras for which a distance estimate is known to exist is presently rather limited. E. Christensen has shown that it contains "most" von Neumann algebras. Distance estimates have proven useful when they have been known to exist. From Proposition 12 it seems that problems concerning distance estimates have the appearance of extreme point problems. Rank-1 operators of norm 1 in the unit ball of the annihilator are extreme points, so in the case of a reflexive algebra the unit ball of the annihilator seems rich in extreme points. These are not in general the only ones as pointed out by finite dimensional examples for which the distance constant is not 1. (That the constant need not be 1 for even a von Neumann algebra in finite dimensions was first pointed out by M.D.Choi.)

QUESTION 7. Does the $||\cdot||_1$-closed convex hull of the set of extreme points in the unit ball of the annihilator of a reflexive operator algebra necessarily have nonempty relative interior?

This would be most interesting in the case of a von Neumann algebra or of a CSL algebra. A positive resolution would not necessarily imply an Arveson distance estimate, but if a suitable characterization of the extreme points were obtained other estimates might arise.

We thank F.Gilfeather and A.Hopenwasser for useful conversations concerning the subject matter presented in this note.

REFERENCES

1. Arveson, W.: Interpolation problems in nest algebras, *J.Functional Analysis* 20(1975), 208-233.

2. Brown, S.; Chevreau, B.; Pearcy, C.: Contractions with rich spectrum have invariant subspaces, *J.Operator Theory* 1(1979), 123-136.

3. Christensen, E.: Extensions of derivations, *J.Functional Analysis* 27(1978), 234-247.

4. Christensen, E.: Extensions of derivations. II, *preprint.*

5. Davidson, K.: Commutative subspace lattices, *Indiana University Math.J.* 27(1978), 479-490.

6. Gilfeather, F.; Larson, D.: Nest-subalgebras of von Neumann algebras: commutants modulo compacts and distance estimates, *J.Operator Theory* 7(1982), 279-302.

7. Lance, E.C.: Cohomology and perturbations of nest algebras, *Proc.London Math.Soc.* (3) 43(1981), 334-356.

D.Larson
Department of Mathematics and Statistics,
University of Nebraska,
Lincoln, Nebraska 68588,
U.S.A.

Supported in part by a grant from the N.S.F.

Added in proof: Since submission of this article we have learned that C.Apostol and J.Langsam have independently answered Question 1 affirmatively.

ON THE DERIVATIONS WITH NORM CLOSED RANGE IN BANACH SPACES

Lu Shijie

The aim of this note is to give some necessary conditions on T under which the inner derivation Δ_T has closed range for spectral operators, nilpotent operators, algebraic operators and some necessary and sufficient conditions for compact operators acting in Banach spaces.

One of the problems on derivations is: "under which conditions on T, the range of the inner derivation Δ_T is norm closed?". Several authors worked on this problem. In 1975 J.G.Stampfli [8] showed that if T is a hyponormal operator in a Hilbert space, then Δ_T has closed range if and only if the spectrum of T is finite. One year later C.Apostol and J.G.Stampfli [2] gave a fairly complete answer for spectral operators, compact operators, weighted shifts and nilpotent operators in Hilbert spaces. Subsequently C.Apostol [3] completely solved this problem in Hilbert spaces. He proved that Δ_T has closed range if and only if the algebra A_T generated by T and I is finite-dimensional and every $A \in A_T$ has closed range. In 1975 J.Anderson and C.Foiaş [1] proved the following result: if T is a scalar operator in the sense of Dunford [4] acting in a Banach space, then Δ_T has closed range if and only if the spectrum of T is finite (in fact, they proved a theorem concerning generalized derivation). It is not known whether Apostol's theorem holds true in Banach spaces. The major difficulties while one tries to solve the problem are the lack of projections onto a given subspace and the absence of a substitute for the non-commutative Weyl-von Neumann theorem of Voiculescu.

In this note we shall give a partial answer to the above question for spectral operators, nilpotent operators, algebraic operators and a complete answer for compact operators. The paper will be divided into three sections. The first section is devoted to preliminaries. In the second section we derive necessary conditions for spectral inner derivations to have closed range. We prove for instance (Theorem 2.6) that the condition "A has closed range for every $A\in A_T$" is necessary for Δ_T to have closed range, in case T is algebraic. In Section 3 we show that Apostol's theorem holds true if T is compact.

1. NOTATION AND PRELIMINARIES

Let $\mathbb{C}$ be the complex field and let X, X_1, X_2 be Banach spaces over $\mathbb{C}$. We shall denote by X^* the conjugate space of X. For every subset $M\subset X$ we denote by $\bar{M}$ the closure of M and by $M^\perp$ the annihilator of M:

$$M^\perp=\{x^*\in X^*,\ x^*(x)=0,\ (\forall)x\in M\}.$$

Let $B(X_1,X_2)$ be the set of all bounded linear operators from X_1 into X_2. If $X_1=X_2=X$, we write $B(X)$ instead of $B(X_1,X_2)$. For an operator $T\in B(X)$, the inner derivation $\Delta_T\in B(B(X))$ is defined as follows:

$$\Delta_T: S\to TS-ST,\qquad (\forall)S\in B(X).$$

We denote the spectrum of T by $\sigma(T)$, the null space of T by ker T, the range of T by R(T). Let $T\in B(X)$ be given and let M be a subspace of X. We shall say that M is an invariant subspace of T, if $TM\subset M$; if $SM\subset M$ for every operator S commuting with T, we shall call M a hyperinvariant subspace of T. Recall that $\gamma(T)$, the reduced minimum modulus of T is

$$\gamma(T)=\inf\{||Tx||,\operatorname{dist}(x,\ker T)\geq 1\}.$$

We know that T has closed range if and only if $\gamma(T)>0$ (Theorem IV.5.2 of [6]).

Put nul T=dimker T (the nullity of T) and let nul'T denote the greatest number $m\leq\infty$ with the following property: for any $\varepsilon>0$, there exists an m-dimensional closed linear manifold M_ε such that $||Tx||\leq\varepsilon||x||$ for every $x\in M_\varepsilon$. We shall call nul'T, the approximate nullity of T. We know that if $T\in B(X)$ has not closed range, then nul'T$=\infty$. (See [6], Theorem IV.5.10).

1.1. LEMMA. *Let* $T\in B(X)$ *be given. If* nul'T$=\infty$, *then for any finite-dimensional subspace* N *and* $\varepsilon>0$, *there exists a vector* $x\in X$ *such that*

$$||x||=\text{dist}(x,N)=1, \quad ||Tx||\le\varepsilon.$$

PROOF. By the definition of nul'T there exists an infinite dimensional subspace M such that $||Tx||\le\varepsilon||x||$ for every $x\varepsilon M$. Since $\dim M>\dim N$, by Lemma IV.2.3 of [6] there exists a vector $x\varepsilon M$ such that

$$||x||=\text{dist}(x,N)=1.$$

Since $x\varepsilon M$, we also have that $||Tx||\le\varepsilon$.

The following lemma is a Banach space version of Proposition 1.7 of [3].

1.2. LEMMA. *Let* $T\varepsilon B(X)$ *be such that* $\sigma(T)=\bigcup_{i=1}^{k}\sigma_i$, $\sigma_i\cap\sigma_j=\emptyset$ for $i\neq j$ *and* σ_i $(i=1,\dots,k)$ *is a closed non-void set. Let* X_i *be the spectral subspace corresponding to* σ_i, $T_i=T|X_i$. *Then* $R(\Delta_T)$ *is closed if and only if all of* $R(\Delta_{T_i})$ *are closed.*

PROOF. Corresponding to the decomposition of $X=X_1\dot{+}\dots\dot{+}X_k$, T has the form

$$\begin{bmatrix} T_1 & & 0 \\ & T_2 & \\ & & \ddots \\ 0 & & & T_k \end{bmatrix}.$$

Suppose $R(\Delta_{T_{i_o}})$ is not closed, then we can choose a sequence $\{S_{i_o}^{(n)}\}_{n=1}^{\infty}$ of operators such that

$$S_{i_o}^{(n)}\varepsilon B(X_{i_o}),\ ||T_{i_o}S_{i_o}^{(n)}-S_{i_o}^{(n)}T_{i_o}||\to 0,\ \text{dist}(S_{i_o}^{(n)},\ker\Delta_{T_{i_o}})\ge 1.$$

Let P_i be the spectral projection corresponding to σ_i (see [6], p. 178) and $A\varepsilon\ker\Delta_T$. Then A commutes with P_i and hence A has the form

$$\begin{bmatrix} A_1 & & 0 \\ & A_2 & \\ & & \ddots \\ 0 & & & A_k \end{bmatrix}$$

where $A_i\varepsilon\ker\Delta_{T_i}$. Putting

$$S^{(n)}=\begin{bmatrix} 0 & \cdots & & & 0 \\ & \ddots & & & \\ & & 0 & & \\ & & & S_{i_o}^{(n)} & \\ & & & & 0 \\ & & & & \ddots \\ 0 & \cdots & & & 0 \end{bmatrix}$$

it is easy to see that

$$\mathrm{dist}(S^{(n)},\ker\Delta_T)\geq 1,\quad ||TS^{(n)}-S^{(n)}T||\to 0$$

and this shows that $\gamma(\Delta_T)=0$.

For the proof of the part "if", we assume that $\gamma(\Delta_T)=0$. Then there exists a sequence $\{S^{(n)}\}_{n=1}^{\infty}$ of operators such that

$$\mathrm{dist}(S^{(n)},\ker\Delta_T)\geq 1,\quad ||TS^{(n)}-S^{(n)}T||\to 0.$$

Corresponding to the decomposition of the space $X=X_1\dot{+}\ldots\dot{+}X_k$, $S^{(n)}$ has the form

$$\begin{bmatrix} S_{11}^{(n)} & \cdots\cdots & S_{1k}^{(n)} \\ & \cdots\cdots & \\ S_{k1}^{(n)} & \cdots\cdots & S_{kk}^{(n)} \end{bmatrix}$$

and hence

$$TS^{(n)}-S^{(n)}T=\begin{bmatrix} T_1S_{11}^{(n)}-S_{11}^{(n)}T_1 & \cdots & T_1S_{1k}^{(n)}-S_{1k}^{(n)}T_k \\ \cdots\cdots\cdots & & \\ T_kS_{k1}^{(n)}-S_{k1}^{(n)}T_1 & \cdots & T_kS_{kk}^{(n)}-S_{kk}^{(n)}T_k \end{bmatrix}.$$

Since

$$||(\Delta_T S^{(n)})x_j||\geq ||P_i||^{-1}||(T_iS_{ij}^{(n)}-S_{ij}^{(n)}T_j)x_j|| \qquad (\forall)x_j\in X_j,$$

we have that

$$||T_iS_{ij}^{(n)}-S_{ij}^{(n)}T_j||\to 0$$

and hence by Rosenblum's theorem [2] we have that

$$||S_{ij}^{(n)}||\to 0 \qquad i\neq j.$$

On the other hand, since $\mathrm{dist}(S^{(n)},\ker\Delta_T)\geq 1$, $||S_{ij}^{(n)}||\to 0$ $(i\neq j)$ will imply that at least one of the following inequalities

$$\mathrm{dist}(S_{ii}^{(n_q)},\ker\Delta_{T_i})\geq (2k||P_i||)^{-1}$$

holds true for some subsequence $\{n_q\}_{q=1}^{\infty}$. It follows that at least one of $\gamma(\Delta_{T_i})$ must be zero. The proof is complete.

2. SPECTRAL DERIVATIONS

In this section we will consider the inner derivation determined by spectral operators, nilpotent operators and algebraic operators (the last two classes of operators are particular classes of spectral operators).

2.1. THEOREM. *Let* $T\varepsilon B(X)$ *be a spectral operator. If* $\sigma(T)$ *is infinite, then* $R(\Delta_T)$ *is not closed.*

PROOF. Let $E(\sigma)$ $(\sigma\varepsilon\Sigma)$ be the spectral resolution of identity for T defined on Σ, where Σ denotes a σ-algebra of subsets of $\mathbb{C}$, and let M be the upper bound of $||E(\sigma)||$. We know that T may be written as $T=N+Q$, where N is the scalar part of T, $N=\int_{\sigma(T)}\lambda E(d\lambda)$ and Q is a quasinilpotent operator commuting with N. Let $\{\lambda_j\}_{j=1}^{\infty}$ be a convergent sequence of distinct points in $\sigma(T)$ with the property:

i) $\sum_{j=1}^{\infty}|\lambda_j-\lambda_{j+1}|^{\alpha}<\infty$ for some $\alpha\varepsilon(0,1)$.

Put $D_j=\{\lambda:|\lambda-\lambda_j|\le\varepsilon_j\}$, where $\{\varepsilon_j\}_{j=1}^{\infty}$ satisfies the following conditions:

ii) $\varepsilon_j\ge\varepsilon_{j+1}$ for $j=1,2,\ldots$;

iii) $\varepsilon_j|\lambda_j-\lambda_{j+1}|^{-1}\to 0$ as $j\to\infty$;

iv) $D_j\cap D_k=\emptyset$ for all $j\neq k$.

If we set $X_j=E(D_j)X$, $X_j^*=E(D_j)^*X^*$ then X_j and X_j^* are invariant under T and T^*, respectively , and the restrictions T_j of T to X_j and T_j^* of T^* to X_j^* may be written as $T_j=\lambda_j I_{X_j}+F_j+Q_j$, $T_j^*=\lambda_j I_{X_j^*}+F_j^*+Q_j^*$, where $||F_j||\le\varepsilon_j M$, $||F_j^*||\le\varepsilon_j M$, Q_j and Q_j^* are quasinilpotent. Choose unit vectors $x_j\varepsilon X_j$, $g_j\varepsilon X_j^*$ such that

$$||Q_jx_j||<\varepsilon_j,\quad ||Q_j^*g_j||<\varepsilon_j,$$

and choose vectors $f_j\varepsilon X_j^*$, $y_j\varepsilon X_j$ such that

$$||f_j||\le M,\quad f_j(x_j)=1,\quad ||y_j||\le M,\quad g_j(y_j)=1-\eta_j,$$

where $\eta_j>0$ is arbitrarily small.

Define $L_j\varepsilon B(X_j, X_{j+1})$ as follows

$$L_jx=|\lambda_j-\lambda_{j+1}|^{\alpha}f_j(x)y_{j+1},\qquad (\forall)\,x\varepsilon X_j.$$

Since $\sigma(T_j)\cap\sigma(T_{j+1})=\emptyset$ for all j, by Rosenblum's theorem [2] we can find $S_j\varepsilon B(X_j,X_{j+1})$ such that

$$T_{j+1}S_j-S_jT_j=L_j.$$

Define

$$V_n=\begin{cases}S_j & \text{on } X_j \quad \text{for } j\leq n;\\ 0 & \text{on } E(\mathbb{C}\setminus\bigcup_{j\leq n} s_j)X.\end{cases}$$

We have that

$$(\Delta_T V_n)x_j'=\begin{cases}|\lambda_j-\lambda_{j+1}|^{\alpha}f_j(x_j')y_{j+1}, & j\leq n,\\ 0, & j>n,\end{cases}\qquad (\forall)\, x_j'\varepsilon X_j$$

and hence

$$||(\Delta_T(V_n)-\Delta_T(V_m))x||\leq M||x||\sum_{j=n+1}^{m}|\lambda_j-\lambda_{j+1}|^{\alpha}\qquad \text{for } n<m.$$

The last inequality shows that $\Delta_T(V_n)$ converges in norm topology to an operator $A\varepsilon B(X)$. We shall show that $A\notin R(\Delta_T)$. To this end, we assume that $A=TW-WT$, for some operator $W\varepsilon B(X)$. Since

$$\begin{aligned}<Ax_j,g_{j+1}>&=|\lambda_j-\lambda_{j+1}|^{\alpha}(1-\eta_{j+1})=<(TW-WT)x_j,\ g_{j+1}>=\\ &=(\lambda_{j+1}-\lambda_j)<Wx_j,\ g_{j+1}>+<Wx_j,F_{j+1}^*g_{j+1}>+\\ &\quad +<Wx_j,Q_{j+1}^*g_{j+1}>-<F_jx_j,\ W^*g_{j+1}>-<Q_jx_j,W^*g_{j+1}>,\end{aligned}$$

we derive

$$|<Wx_j,g_{j+1}>|\geq(1-\eta_{j+1})|\lambda_j-\lambda_{j+1}|^{\alpha-1}-4M\varepsilon_j|\lambda_j-\lambda_{j+1}|^{-1}||W||\to\infty.$$

Hence W is not bounded and the proof is complete.

Let T be a nilpotent operator of order n. Define the quotient spaces

$$[X]_k=\ker T^{k+1}/\ker T^k,\qquad k=0,1,\dots,n-1,$$

and the operators $[T]_k\varepsilon B([X]_k,[X]_{k-1})$ as follows:

$$[T]_k[x]_k=[Tx]_{k-1},\quad x\varepsilon[x]_k\varepsilon[X]_k,\quad k=1,2,\dots,n-1.$$

Clearly we have that

$$([X]_k)^*=\overline{R(T^{*k})}/\overline{R(T^{*k+1})},\quad ([T]_k)^*\varepsilon B(([X]_{k-1})^*,([X]_k)^*).$$

For simplicity we shall write $[X]_k^*$ for $([X]_k)^*$, $[T]_k^*$ for $([T]_k)^*$.

2.2. LEMMA. *Let* T *be a nilpotent operator of order* n *and let* $[T]_k$ *be defined as above. Then the following implications hold true:*

i) $R(T^{k_o})=\overline{R(T^{k_o})}\Longrightarrow\gamma([T]_{k_o})>0,$

ii) $R(T^k)=\overline{R(T^k)},\ 1\leq k\leq n-1\Longleftrightarrow \min_{1\leq k\leq n-1}\gamma([T]_k)>0.$

PROOF. i) If $R(T^{k_o})$ is closed, then there exists a positive number δ such that

$$||T^{k_o}x||\geq\delta\,\mathrm{dist}(x,\ker T^{k_o}).$$

Since

$$||[T]_1\ldots[T]_{k_o}[x]_{k_o}||=||T^{k_o}x||\geq\delta||[x]_{k_o}||,\qquad (\forall)[x]_{k_o}\varepsilon[X]_{k_o},$$

we have

$$||[T]_{k_o}[x]_{k_o}||\geq||[T]_1\ldots[T]_{k_o-1}||^{-1}\delta||[x]_{k_o}||.$$

This shows that $\gamma([T]_k)>0$.

ii) The part "only if" follows from i). In order to prove the part "if", we suppose $\min_{1\leq k\leq n-1}\gamma([T]_k)>0$ and $R(T^{k_o})\neq\overline{R(T^{k_o})}$ for some k_o, then derive a contradiction. Let k_o be the maximal natural number with the property: $R(T^k)\neq\overline{R(T^k)}$. Hence there exists a sequence $\{x_j\}_{j=1}^{\infty}$ such that

$$\mathrm{dist}(x_j,\ker T^{k_o})=1,\qquad ||T^{k_o}x_j||\to 0.$$

Since T^{k_o+1} has closed range and $||T^{k_o+1}x_j||\to 0$, we may assume that $x_j\varepsilon\ker T^{k_o+1}$. Further using the relation

$$T^{k_o}x_j=[T]_1\ldots[T]_{k_o}[x_j]_{k_o}$$

and the fact that $[T]_k$, $1\leq k\leq n-1$, is bounded from below, we derive

$$\lim_{j\to\infty}||[x_j]_{k_o}||=\lim_{j\to\infty}\mathrm{dist}(x_j,\ker T^{k_o})=0,$$

which is a contradiction.

For any $x\varepsilon X$ and $x^*\varepsilon X^*$, define the rank-one operator $x\otimes x^*\varepsilon B(X)$ by the equation

$$(x\otimes x^*)z=\langle z,x^*\rangle x,\qquad (\forall)z\varepsilon X.$$

2.3. LEMMA. *Suppose* $\ker T\neq\ker T^2$ *and* $R([T]_1)\neq\overline{R([T]_1)}$. *Then there exists a sequence* $\{S_n\}_{n=1}^{\infty}\subset B(X)$ *such that* rank $S_n=1$,

$$\mathrm{dist}(S_n,\ker\Delta_T)\geq 1,\quad \lim_{n\to\infty}||\Delta_T S_n||=0.$$

PROOF. Since by [6], Theorem IV.5.13, we know that $[T]_1^*$ has not closed range. Hence we can find $\{x_n^*\}_{n=1}^{\infty}\subset X^*$ such that

$$\mathrm{dist}(x_n^*,(\ker T)^{\perp})=1>||x_n^*||-\frac{1}{n},\ \lim_{n\to\infty}\mathrm{dist}(T^*x_n^*,(\ker T^2)^{\perp})=0.$$

Since $(\ker T^2)^{\perp}=\overline{T^{*2}X^*}$, we can find $y_n^*\in X^*$ such that

$$\lim_{n\to\infty}||T^*x_n^*-T^{*2}y_n^*||=0.$$

Let $\{x_n\}_{n=1}^{\infty}\subset \ker T^2$ be such that (via $R([T]_1)\neq\overline{R([T]_1)}$)

$$\mathrm{dist}(x_n,\ker T)=1>||x_n||-\frac{1}{n},\quad \lim_{n\to\infty}||Tx_n||\cdot||x_n^*-T^*y_n^*||=0.$$

If we define $S_n\in B(X)$ by the equation

$$S_n=x_n\otimes(x_n^*-T^*y_n^*)$$

we have

$$\lim_{n\to\infty}||\Delta_T S_n||=\lim_{n\to\infty}||Tx_n\otimes(x_n^*-T^*y_n^*)-x_n\otimes(T^*x_n^*-T^{*2}y_n^*)||=0.$$

Since for every $A\in\ker\Delta_T$ we have $A\ker T\subset\ker T$, we derive

$$\begin{aligned}||S_n-A||&\geq\sup\{\mathrm{dist}(S_ny,\ker T):y\in\ker T,\ ||y||=1\}=\\&=\sup\{\mathrm{dist}(x_n^*(y)x_n,\ker T):y\in\ker T,||y||=1\}=\\&=\sup\{|x_n^*(y)|:y\in\ker T,||y||=1\}=\\&=\mathrm{dist}(x_n^*,(\ker T)^{\perp})=1,\end{aligned}$$

thus $\mathrm{dist}(S_n,\ker\Delta_T)\geq 1$. This shows that $R(\Delta_T)$ is not closed.

2.4. PROPOSITION. *Let* $m\geq 2$ *be given and let* T *be a nilpotent operator of order* m, *such that* $\min_{1\leq k\leq m-1}\gamma([T]_k)=0$. *Then there exists a sequence* $\{S_n\}_{n=1}^{\infty}\subset B(X)$ *such that*

$$\mathrm{rank}S_n\leq r_m,\ \mathrm{dist}(S_n,\ker\Delta_T)\geq 1,\ \lim_{n\to\infty}||\Delta_T S_n||=0,$$

where r_m *depends only on* m.

PROOF. We proceed by induction. Since for $m=2$ we can apply Lemma 2.3, we shall assume $m>2$. If $\gamma([T]_1)=0$, we apply again Lemma 2.3, thus we shall also assume $\gamma([T]_1)>0$. Let $\tilde{T}$ denote the quotient operator determined by T in the quotient space $X/\ker T$. It is easy to see that $\tilde{T}$ is a nilpotent operator of order $m-1$ and

$$\min_{1\leq k\leq m-2}\gamma([\tilde{T}]_k)=\min_{1\leq k\leq m-2}\gamma([T]_{k+1})=0.$$

By induction hypothesis we can find $\{A_n\}_{n=1}^{\infty}\subset B(X/\ker T)$ such that

$$\mathrm{rank}A_n\leq r_{m-1},\mathrm{dist}(A_n,\ker\Delta_{\tilde{T}})\geq 1,\ \lim_{n\to\infty}||\Delta_{\tilde{T}}A_n||=0.$$

We may suppose that we have

$$A_n=\sum_{j=1}^{r_{m-1}}\tilde{x}_{n,j}\otimes x_{n,j}^*,\quad \Delta_T(\sum_{j=1}^{r_{m-1}}x_{n,j}\otimes x_{n,j}^*)=\sum_{j=1}^{r'_{m-1}}y_{n,j}\otimes x_{n,j}^*$$

where $r'_{m-1}\leq 2r_{m-1}$, $x_{n,j}^*\in(\ker T)^{\perp}$, $||x_{n,j}^*||=1$, $\lim_{n\to\infty}\mathrm{dist}(y_{n,j},\ker T)=0$.

Choose $z_{n,j} \varepsilon \ker T$, $z^*_{n,j} \varepsilon X^*$ such that $\lim_{n\to\infty} ||y_{n,j} - z_{n,j}|| = 0$, $\lim_{n\to\infty} ||x^*_{n,j} - T^* z^*_{n,j}|| = 0$ (via $(\ker T)^{\perp} = \overline{T^* X^*}$) and put

$$S_n = \sum_{j=1}^{r_{m-1}} x_{n,j} \otimes x^*_{n,j} + \sum_{j=1}^{r'_{m-1}} z_{n,j} \otimes z^*_{n,j}.$$

Then plainly rank $S_n \leq 3r_{m-1}$, $\lim_{n\to\infty} ||\Delta_T S_n|| = 0$ and $\operatorname{dist}(S_n, \ker \Delta_T) \geq$ $\geq \operatorname{dist}(A_n, \ker \Delta_{\tilde{T}}) \geq 1$.

2.5. COROLLARY. *Let* T *be a nilpotent operator of order* n *such that* $R(\Delta_T)$ *is closed. Then* $R(T^k)$ *is closed for every* $1 \leq k \leq n-1$.

PROOF. If the conclusion of our corollary is false, then using Lemma 2.2(ii) we can find k_o such that $\gamma([T]_{k_o}) = 0$. But applying Proposition 2.4 we derive $\gamma(\Delta_T) = 0$, a contradiction.

2.6. THEOREM. *Let* $T \varepsilon B(X)$ *be an algebraic operator such that* $R(\Delta_T)$ *is closed. Then* $R(p(T))$ *is closed for every polynomial* p.

PROOF. Since T is algebraic, $\sigma(T)$ is a finite set. Suppose $\sigma(T) = \{\lambda_1, \ldots, \lambda_k\}$ and put $\sigma_i = \{\lambda_i\}$. Let X_i, T_i be defined as in Lemma 1.2. By Lemma 1.2 we have that $R(\Delta_{T_i})$ $(i=1,\ldots,k)$ is closed and hence $R(\Delta_{T_i - \lambda_i})$ is closed. But because $T_i - \lambda_i$ is nilpotent, and by Corollary 2.5, $R((T_i - \lambda_i)^k)$ is closed, $k=1,2,\ldots$, the conclusion of our theorem follows.

3. COMPACT DERIVATIONS

In this section we shall prove the following:

3.1. THEOREM. *Let* $T \varepsilon B(X)$ *be compact. Then* $R(\Delta_T)$ *is closed if and only if* $R(T)$ *is closed.*

We shall need the following lemmas.

3.2. LEMMA. *Let* $T \varepsilon B(X)$ *be compact. Then there exists a sequence* $\{Q_k\}_{k=1}^{\infty}$ *of one-dimensional, norm* 1 *projections, such that* $Q_i Q_j = 0$ *for* $i > j$ *and* $||TQ_k - Q_k T|| \to 0$ *as* $k \to \infty$.

PROOF. By Theorem IV.5.10 of [6] we have that $\text{nul}' T = \infty$. By induction, using Lemma 1.1, we can construct four sequences $\{x_k\}_{k=1}^{\infty}$, $\{x'_k\}_{k=1}^{\infty}$, $\{x^*_k\}_{k=1}^{\infty}$ and $\{N_k\}_{k=1}^{\infty}$ such that

$$x_k,\ x'_k \varepsilon X,\ N_k = \operatorname{sp}\{x_1, Tx'_1, \ldots, x_k, Tx'_k\},$$

$$||x_k||=\text{dist}(x_k,N_{k-1})=1,\ ||Tx_k||<\frac{1}{k},\qquad N_o=\{0\},$$
$$x_k^*\varepsilon N_{k-1}^{\perp},\ ||x_k^*||=x_k^*(x_k)=1,$$
$$||x_k'||=1,\ |T^*x_k^*(x_k')|>||T^*x_k^*||-\frac{1}{k}.$$

We shall show that there exists a subsequence $\{x^*_{n_k}\}_{k=1}^{\infty}$ such that $||T^*x^*_{n_k}||\to 0$. In fact, since $\{||x_n^*||\}_{n=1}^{\infty}$ is bounded and T^* is compact, there exists a subsequence $\{x^*_{n_k}\}_{k=1}^{\infty}$ such that $\{T^*x^*_{n_k}\}_{k=1}^{\infty}$ converges. For k>j, we have that

$$||T^*x^*_{n_k}-T^*x^*_{n_j}||\geq||T^*x^*_{n_k}(x'_{n_j})-T^*x^*_{n_j}(x'_{n_j})||>||T^*x^*_{n_j}||-\frac{1}{n_j}.$$

Hence we have that $||T^*x^*_{n_j}||\to 0$. Define

$$Q_k=x_{n_k}\otimes x^*_{n_k}.$$

Obviously we have that $||TQ_k-Q_kT||\to 0$.

3.3. LEMMA. *Suppose* $T\varepsilon B(X)$ *is compact and has not closed range. If* nul$T=\infty$, *then* Δ_T *has not closed range.*

PROOF. Since nul$T=\infty$, we can choose $\{y_n\}_{n=1}^{\infty}\subset \ker T$, $\{f_n\}_{n=1}^{\infty}\subset X^*$ such that

$$||y_n||=1,\quad \text{dist}(y_n,\ sp\{y_i\}_{i=1}^{n-1})=1,$$
$$||f_n||=1,\quad f_n(y_m)=\delta_{nm}\ \text{for}\ m\leq n.$$

Since $\{||f_n||\}_{n=1}^{\infty}$ is bounded, by Theorem V.4.2 of [5], we may assume that $\{f_n\}_{n=1}^{\infty}$ is weak*-convergent to an element f. Clearly we have that $f(y_n)=0$ for all n. Setting $g_n=f_n-f$, we have that $g_n(y_n)=1$, $g_n\xrightarrow{w^*}0$ and $\{||g_n||\}_{n=1}^{\infty}$ is bounded. By Theorem VI.5.6 of [5] we have that $||T^*g_n||\to 0$. Since R(T) is not closed, using Theorem IV.5.2 of [6], we can find a sequence $\{x_n\}_{n=1}^{\infty}$ such that

$$||x_n||<1+\frac{1}{n},\ \text{dist}(x_n,\ker T)\geq 1,\ ||Tx_n||\to 0.$$

Define

$$S_n=x_n\otimes g_n.$$

We have that

$$\lim_{n\to\infty}||\Delta_T S_n||=\lim_{n\to\infty}||Tx_n\otimes g_n-x_n\otimes T^*g_n||=0.$$

On the other hand

$$\text{dist}(S_n,\ker\Delta_T)=\inf_{A\varepsilon\ker\Delta_T}||S_n-A||\geq$$
$$\geq\inf_{A\varepsilon\ker\Delta_T}||(S_n-A)y_n||\geq\text{dist}(x_n,\ker T)\geq 1.$$

This proves that $\gamma(\Delta_T)=0$.

3.4. LEMMA. *Suppose* $T\varepsilon B(X)$ *is compact and has not closed range. If there exists a sequence* $\{P_n\}_{n=1}^{\infty}$ *of one-dimensional projections such that* $P_iX\neq P_jX$ *for* $i\neq j$, $||P_n||\to 1$, $TP_n=P_nT=\lambda_nP_n$, *then* Δ_T *has not closed range.*

PROOF. Clearly every λ_n is an eigenvalue of T. Let N_{λ_T} be the eigenspace corresponding to $\{\lambda_n\}$, then N_{λ_n} is a hyperinvariant subspace of T. If $\lambda_n=0$ for infinitely many n's, then by Lemma 3.3 we derive that Δ_T has not closed range, thus we may assume that all of λ_n are not zero. Since T is compact, we have that $\dim N_{\lambda_n}<\infty$. By Lemma 1.1, there exists a vector x_n such that

$$||x_n||=\operatorname{dist}(x_n,N_{\lambda_n})=1, \qquad ||Tx_n||<\frac{1}{n}.$$

Let y_n be a unit vector in P_nX. If we choose $x_n^*\varepsilon X^*$ such that $||x_n^*||=x_n^*(y_n)=1$ and if we put $f_n=P_n^*x_n^*$, we have $f_n(y_n)=1$, $T^*f_n=\lambda_nf_n$ and $\{||f_n||\}_{n=1}^{\infty}$ is bounded. Define

$$Q_n=x_n\otimes f_n.$$

Since $AN_{\lambda_n}\subset N_{\lambda_n}$ for every $A\varepsilon\ker\Delta_T$, we have that

$$\operatorname{dist}(Q_n,\ker\Delta_T)\geq \inf_{A\varepsilon\ker\Delta_T}||Q_n-A||\geq \inf_{A\varepsilon\ker\Delta_T}||(Q_n-A)y_n||=$$

$$=\inf_{A\varepsilon\ker\Delta_T}||x_n-Ay_n||\geq\operatorname{dist}(x_n,N_{\lambda_n})=1.$$

On the other hand, we have that

$$||TQ_n-Q_nT||=||Tx_n\otimes f_n-x_n\otimes T^*f_n||\leq$$
$$\leq||Tx_n||\cdot||P_n||+||T^*f_n||\to 0,$$

and this shows that $\gamma(\Delta_T)=0$.

PROOF OF THEOREM 3.1. The proof of the part "if" is similar with the corresponding part of Theorem 2 of [2]. To prove the part "only if", we assume that R(T) is not closed and we prove that $R(\Delta_T)$ is not closed.

By Lemma 3.2 there exists a sequence $\{Q_k\}_{k=1}^{\infty}$ of one-dimensional projections such that $Q_i\neq Q_j$ for $i\neq j$ and $||TQ_k-Q_kT||\to 0$ as $k\to\infty$. If $\overline{\lim_{k\to\infty}}\operatorname{dist}(Q_k,\ker\Delta_T)>0$, then $R(\Delta_T)$ is not closed. So we assume that exists a sequence $\{A_k\}_{k=1}^{\infty}$ such that $A_k\varepsilon\ker\Delta_T$ and $||Q_k-A_k||\to 0$.

By standard functional calculus arguments we can construct projections $P_k \varepsilon \ker \Delta_T$ such that $||P_k - Q_k|| \to 0$. By Problem III.3.21 of [6], $\mathrm{rank} P_k = \mathrm{rank} Q_k = 1$ and hence we have

$$TP_k = P_k T = \lambda_k P_k .$$

We may suppose (eventually discarding a finite number of P_i's) that we have $P_i X \neq P_j X$ for $i \neq j$. Indeed, otherwise we can find $i_k, j_k \geq 1$ such that $\lim_{k\to\infty} i_k = \infty$, $P_{i_k + j_k} X = P_{i_k} X$ and if $x_k \varepsilon P_{i_k} X$, $||x_k|| = 1$ we have

$$0 = \lim_{k\to\infty} ||Q_{i_k + j_k} Q_{i_k} x_k|| = \lim_{k\to\infty} ||P_{i_k + j_k} P_{i_k} x_k|| = \lim_{k\to\infty} ||x_k|| = 1,$$

which is a contradiction. Now applying Lemma 3.4 we derive $\gamma(\Delta_T) = 0$ and this concludes the proof.

GENERAL REMARKS.

To decide whether Δ_T has closed range if and only if T is algebraic and p(T) has closed range for every polynomial p, one has to answer the following questions:

QUESTION 1. Does Δ_T have closed range if T has closed range and $T^2 = 0$?

QUESTION 2. Does T have closed range if Δ_T has closed range?

QUESTION 3. Is T necessarily a nilpotent operator if Δ_T has closed range and $\sigma(T) = \{0\}$?

It seems probably that a positive answer to the above questions will provide a positive solution to the general problem. If Question 2 and Question 3 only are positively answered then the necessary part will presumably follow.

Acknowledgement. The author is grateful to Professor C.Apostol for his many useful suggestions.

REFERENCES

1. Anderson, J.H.; Foiaş, C.: Properties normal operators share with normal derivations and related operators, *Pacific J.Math.* 61(1975), 313-326.

2. Apostol, C.; Stampfli, J.G.: On derivation ranges, *Indiana Univ.Math.J.*25(1976), 857-869.

3. Apostol, C.: Inner derivations with closed range, *Rev.Roumaine Math.Pures Appl.*20(1976), 249-265.

4. Dunford, N.: Spectral operators, *Pacific J.Math.*4(1954), 321-354.

5. Dunford, N.; Schwartz, J.T.: *Linear operators.I.*, Interscience Publishers, New-York, 1958.

6. Kato, T.: *Perturbation theory for linear operators*, Springer-Verlag, Berlin, 1966.

7. Rosenblum, M.A.: On the operator equation BX-XB=Q, *Duke Math.J.*23(1956), 263-270.

8. Stampfli, J.G.: On the range of a hyponormal derivation, *Proc.Amer.Math.Soc.*52(1975), 117-120.

Lu Shijie
Department of Mathematics,
University of Bucharest,
Bucharest, Romania,

and

Department of Mathematics,
University of Nanjing,
Nanjing,China.

ON BOOLEAN ALGEBRAS OF PROJECTIONS AND PRESPECTRAL OPERATORS

B.Nagy

1. INTRODUCTION

Complete and σ-complete Boolean algebras of projections in a complex Banach space were studied first by Bade [1] (see also [3; XVII.3]). The purpose of this paper is to find the appropriate extensions of several of his results to the more general case of G-complete and G-σ-complete Boolean algebras of projections, where G is a total linear manifold in the dual of the underlying Banach space. We shall prove e.g. that a Boolean algebra of projections is G-σ-complete if and only if it coincides with the range of a spectral measure of class G (Theorem 2), and we shall give a sufficient condition ensuring that the uniformly closed operator algebra generated by a G-σ-complete Boolean algebra B of projections coincides with the first commutant of B (Theorem 3). The new techniques will include the application of certain weak topologies and some of the duality theory of paired linear spaces as well as an idea due to Palmer [5].

As σ-complete Boolean algebras of projections are closely connected with the theory of spectral operators in the sense of Dunford, a similar connection exists between G-σ-complete Boolean algebras of projections and prespectral operators of class G (for the latter see Dowson [2]). Exploiting some of the methods of Section 2, in Section 4 we study when an operator T is prespectral of class G, provided its adjoint T^* is prespectral of class X. In Section 3 we discuss an example showing some basic difference between the theories of (norm-) σ-complete and G-σ-complete Boolean algebras of projections.

Now we shall fix some notations and conventions that will be used throughout the paper. If X and G are paired linear spaces with a fixed bilinear functional on their product then, according to convenience, the latter will be denoted in one of the following ways:

$$(x,g)=g(x)=gx=(g,x) \qquad (x\varepsilon X,\ g\varepsilon G).$$

The weak (X,G) topology of X determined by this pairing (see e.g.,[4; p.138]) will be denoted by $w(X,G)$. If Y and H are also paired linear spaces then $C(w(X,G),\ w(Y,H))$ will denote the family of all continuous mappings of the linear topological space $(X,w(X,G))$ into the linear topological space $(Y,w(Y,H))$. If T is a linear operator belonging to this family, then its dual, belonging to $C(w(H,Y),w(G,X))$, will be denoted by T'. The term adjoint will be reserved for the Banach space dual T^* of an operator T belonging to $B(X)$, the Banach algebra of bounded linear operators in the complex Banach space X. If H is a set in the Banach space X, then span H=sp H denote the linear hull of H and $\overline{\text{sp}}\ H$ denotes the norm-closure of sp H. If b and c are sets, we often write bc for $b \cap c$. Finally, let E be a spectral measure of class (M,G), where M is a σ-algebra of subsets of a set S. Then $EB(S,M)$ will denote the algebra of all complex-valued M-measurable E-essentially bounded functions on S.

2. ON BOOLEAN ALGEBRAS OF PROJECTIONS

Let B be a Boolean algebra of projections in the complex Banach space X, with respect to the operations

$$E_1 \vee E_2=E_1+E_2-E_1E_2,\quad E_1 \wedge E_2=E_1E_2 \qquad (E_1,E_2\varepsilon B).$$

Let G be a total linear manifold in the dual space X^*. Let B^c denote the commutant of B in $B(X)$ and, following an idea of Palmer (cf.[5; pp.407-408]), let

$$B^*G=\text{span}\{T^*g;\ T\varepsilon B^c,\ g\varepsilon G\}.$$

Then $G \subset B^*G \subset X^*$ and $B^*B^*G=B^*G$. Any operator U in B^c is continuous as a mapping of X endowed with the weak topology $w(X,B^*G)$ into itself; in notation $U\varepsilon C(w(X,B^*G),\ w(X,B^*G))$. Further, a net $\{E_a\}$ of projections in B converges to $E\varepsilon B$ in the G operator topology, i.e. $\lim_a gE_ax=gEx$ for all $x\varepsilon X$, $g\varepsilon G$, if and only if $\{E_a\}$ converges to E in the B^*G operator topology. We shall denote this either by $G\text{-}\lim_a E_a=E$ or by $G\text{-}\lim_a E_ax=Ex$ $(x\varepsilon X)$.

DEFINITION 1. A net $\{E_a;\ a\in A\}$ of linear operators mapping the linear topological space (X,T_x) into the linear topological space (Y,T_y) is said *to become small on small sets* (with respect to the topologies T_x-T_y) if for every neighborhood U of $0\in Y$ there is a neighborhood V of $0\in X$ such that for every $x\in V$ there is $a_o\in A$ with the property that $a\geq a_o$ implies $E_a x\in U$.

As a trivial example, an equicontinuous net of linear operators becomes small on small sets.

DEFINITION 2. A Boolean algebra B of projections in X is called *G-complete* (*G-σ-complete*) if

1°. B is complete (σ- complete) as an abstract Boolean algebra;

2°. For every set (sequence) B_o in B

$$(\bigvee(E;E\in B_o))X=B^*G\text{-sp}\ (EX;\ E\in B_o),$$
$$(\bigwedge(E;\ E\in B_o))X=\bigcap(EX;\ E\in B_o),$$

where B^*G-sp (·) denotes the closure in the $w(X,B^*G)$ topology of the linear span of the set standing in the parentheses;

3°. Each monotonic net (sequence) B_o in B becomes small on small sets with respect to the $w(X,B^*G)$-$w(X,B^*G)$ topologies.

THEOREM 1. (a) *Let* B *be a* G-*complete* (G-σ-*complete*) *Boolean algebra of projections in* X, *and let* $\{E_a\}$ *be a monotonic net (sequence) in* B. *If* $\{E_a\}$ *is increasing, then*

$$G\text{-}\lim_a E_a=\bigvee_a E_a,$$

whereas if $\{E_a\}$ *is decreasing, then*

$$G\text{-}\lim_a E_a=\bigwedge_a E_a.$$

(b) *Conversely, if every monotonic increasing net (sequence) of elements of a Boolean algebra* B *of projections converges in the* G *operator topology to some element of* B, *then* B *is* G-*complete* (G-σ-*complete*).

PROOF. (a) Let B be G-complete, let $\{E_a;a\in A\}$ be a monotonic increasing net, and let $E=\bigvee_a E_a$, $\varepsilon>0$, $x\in X$, $g\in G$. Since $\{E_a\}$ becomes small on small sets with respect to the $w(X,B^*G)$-$w(X,B^*G)$ topologies, for the neighborhood

$$U=\{r\in X;\ |g(r)|<\varepsilon\}$$

there is a neighborhood V of 0 such that for every z in V there is $a(U,z)\varepsilon A$ for which $a\geq a(U,z)$ implies

$$|g(E_a z)|<\varepsilon.$$

Let $W=U\cap V$. Since W is a neighborhood of 0 and $EX=B^*G\text{-sp }(E_aX;\ a\varepsilon A)$, there are $a_o\varepsilon A$ and $y\varepsilon E_{a_o}X$ such that $Ex-y\varepsilon W$. Hence $|g(y-Ex)|<\varepsilon$. Further, for each $a\geq a(U,Ex-y)$ we have $|g(E_a(Ex-y))|<\varepsilon$. If $a_1\geq a(U,Ex-y)$ and $a_1\geq a_o$, then for every $a\geq a_1$ we have $y=E_{a_o}y=E_a y$, hence

$$|gE_a x-gEx|\leq|gE_a(Ex-y)|+|g(y-Ex)|<2\varepsilon.$$

Therefore $G\text{-}\lim_a E_a=E$. The dual statement concerning decreasing nets reduces to the preceding one, and the proof for a G-σ-complete Boolean algebra B is similar.

(b) Let $\{E\}$ be a set in B. Let $\{F_a;\ a\varepsilon A\}$ be the net of finite unions of elements of $\{E\}$, ordered by the natural ordering of projections. Clearly, $F=\bigvee\{E\}$ if and only if $F=\bigvee\{F_a\}$. We note that if $P\varepsilon B$, then PX is a $w(X,B^*G)$-closed subspace. Indeed, if $B^*G\text{-}\lim_b x_b=x$ and $Px_b=x_b$, then $P\varepsilon C(w(X,B^*G),\ w(X,B^*G))$ implies that $Px=x$. Hence if $E_1,\ldots,E_n\varepsilon B$, then

$$(E_1\vee\ldots\vee E_n)X=\overline{\text{sp}}\ (E_iX;\ i=1,\ldots,n)=B^*G\text{-sp}\ (E_iX;i=1,\ldots,n).$$

Now we shall show that $\bigvee\{F_a\}$ exists and satisfies the corresponding equality in 2° of Definition 2. By assumption, there is $F\varepsilon B$ such that

$$F=G\text{-}\lim_a F_a=B^*G\text{-}\lim_a F_a.$$

For any $a\varepsilon A$, $g\varepsilon B^*G$ and $x\varepsilon X$ we have

$$gFF_a x=\lim_b gF_bF_a x=gF_a x.$$

Since B^*G is total, $F_a\leq F$. If there is $F'\varepsilon B$ such that $F_a\leq F'$ for every $a\varepsilon A$, then

$$gFF'x=\lim_a gF_aF'x=gFx \qquad (x\varepsilon X,\ g\varepsilon B^*G).$$

Hence $F\leq F'$ and therefore $F=\bigvee\{F_a\}$. Since $Fx=B^*G\text{-}\lim_a F_a x$ $(x\varepsilon X)$, therefore $FX\subset B^*G\text{-sp}\ (F_aX;\ a\varepsilon A)$. On the other hand, $F\varepsilon B$ implies that FX is $w(X,B^*G)$-closed. Hence we obtain the converse containment relation, and therefore

$$FX=B^*G\text{-sp}\ (F_aX;\ a\varepsilon A).$$

From the assumptions it simply follows that any decreasing net converges in the B^*G operator topology to some element of B. Hence we obtain, similarly as above that $\bigwedge\{E\}$ exists and $(\bigwedge\{E\})X=\bigcap(EX)$.

Now let $\{E_a;\ a\varepsilon A\}$ be an arbitrary monotonic net in B. By assumption, there is $E\varepsilon B$ such that

(1) $$B^*G\text{-}\lim_a E_a = E.$$

Since E is $w(X,B^*G)$-$w(X,B^*G)$ continuous, for any neighborhood

$$U=\{y\varepsilon X;\ |g_i(y)|<\varepsilon,\ g_i\varepsilon B^*G \text{ for } i=1,\dots,n\}$$

of 0 there is a neighborhood V of 0 such that for $x\varepsilon V$ we have

$$|g_i(Ex)|<\varepsilon \qquad (i=1,\dots,n).$$

By (1), for any $x\varepsilon V$ there is $a_o=a_o(U,x)$ such that $a\geq a_o$ implies

$$|g_i(E_a x)|<2\varepsilon \qquad \text{for all } i=1,\dots,n,$$

i.e., $E_a x\varepsilon U$. Therefore $\{E_a\}$ satisfies 3° of Definition 2, so B is G-complete. The proof for the G-σ-complete case is similar.

REMARK 1. The preceding theorem generalizes a result of Bade (cf.[3; XVII.3.4]). Indeed, if B is X^*-complete (X^*-σ-complete) in the sense of Definition 2, then in 2° of Definition 2 the B^*X^*-, i.e. the X^*-span is identical with the strong closure of the span, so B is complete (σ-complete) in the sense of Bade [3;XVII.3.1]. Conversely, [3;XVII.3.4] and Theorem 1 here show that if B is complete (σ-complete), then B is G-complete (G-σ--complete) for any total set G in X^*. So B is complete (σ-complete) if and only if B is X^*-complete (X^*-σ-complete).

REMARK 2. If B is G-σ-complete, then B is σ-complete as an abstract Boolean algebra. Therefore, by[3;XVII.3.3], there is $K>0$ such that

$$|E|\leq K \qquad \text{for all } E\varepsilon B.$$

Now we turn to the representation problem of G-σ-complete Boolean algebras of projections.

Let M be a σ-algebra of subsets of an arbitrary set H, and let G be a total linear manifold in X^*. We recall that a mapping E of M into B(X) is called a *spectral measure of class* (M,G) if

(i) $E(H)=I$;

(ii) $E(H\setminus b)=I-E(b)$ for all $b\varepsilon M$;

(iii) $E(b_1b_2)=E(b_1)E(b_2)$ for all $b_1,b_2\varepsilon M$;

(iv) For every x in X and g in G the set function $gE(\cdot)x$ is countably additive on M.

Note that in the definition of a spectral measure of class (M,G) it is usually assumed that, in addition to (i)-(iv), the following property holds (see, e.g.,[2; p.119]):

$$\text{(v)} \quad \sup_{b\in M} |E(b)|<\infty .$$

We shall see immediately that condition (v) is redundant.

LEMMA 1. *If a Boolean algebra* B *of projections in* X *is identical with the range of a spectral measure* E *of class* (M,G), *then* B *is* G-σ-*complete.*

PROOF. Let $\{E_n\}$ be an increasing sequence in B. By assumption, $E_n=E(e_n)$ $(e_n\in M)$, and

$$E(e_n\setminus e_{n+1})=E(e_n)-E(e_ne_{n+1})=E_n-E_nE_{n+1}=0.$$

Hence there is an increasing sequence $\{f_n\}$ in M such that $E_n=E(f_n)$, therefore the limit

$$G\text{-}\lim_n E_n=G\text{-}\lim_n E(f_n)=E(\bigcup_n f_n)$$

exists and belongs to B. By Theorem 1, B is G-σ-complete.

COROLLARY. *In the definition of a spectral measure of class* (M,G) *condition* (v) *is redundant.*

LEMMA 2. *Let* U(B) *be the uniformly closed operator algebra generated by a* G-σ-*complete Boolean algebra* B *of projections in* X. *Then* U(B) *is equivalent to* C(S), *the algebra of continuous functions on the space* S *of the maximal ideals of* U(B) *endowed with the Gelfand topology. Each homeomorphic isomorphism* $T:C(S)\to U(B)$ *uniquely determines a spectral measure E of class* (M,G), *where* M *is the* σ-*algebra of the Borel subsets of* S , *such that*

$$T(f)=\int_S f(s)E(ds) \qquad (f\in C(S)),$$

and the set function $gE(\cdot)x$ *is regular for any* $x\in X$, $g\in G$. *Further, the range of* E *is exactly* B.

PROOF. By Remark 2 after Theorem 1, B is bounded, therefore [3; XVII. 2.1] yields the first statement. By[3;XVII.2.4], for T there is a unique spectral measure A of class (M,X) in X^* such that the measure $xA(\cdot)X^*$ is regular for every x in X and x^* in X^*, and

$$T(f)^*=\int_S f(s)A(ds) \qquad (f\in C(S)).$$

Let $Q=B^*G$ and let N denote the family of all elements $e\in M$ for which there is $E=E(e)\varepsilon B$ such that

$$A(e)g=E(e)^*g \qquad \text{for all } g\varepsilon Q.$$

Since $B^*Q=Q$, for $e\varepsilon N$ the operator $A(e)$ maps Q into Q. Let the linear spaces X and Q be given their natural pairing: $(x,g)=g(x)$ for $x\varepsilon X$ and $g\varepsilon Q$. Then the transformations $E(e):X\to X$ and $A(e)|Q:$ $:Q\to Q$ are duals with respect to the pairing above. We shall express this fact by writing

$$(2) \qquad A(e)=E(e)' \qquad (e\varepsilon N).$$

Since Q is total in X^*, the operator $E(e)$ is uniquely determined.

Since B is a Boolean algebra and A is a spectral measure, it is easily seen that N is an algebra of sets. Let $\{e_n\}$ be an increasing sequence in N. Then $A(e_{n+1})A(e_n)=A(e_n)$ and, by the duality above, $E(e_{n+1})E(e_n)=E(e_n)$. Thus the sequence of projections $\{E(e_n)\}$ is increasing and, since B is G-σ-complete, $E=Q$-$\lim_n E(e_n)$ exists and belongs to B. Therefore, for every $x\varepsilon X$, $g\varepsilon Q$

$$xA(\bigcup_n e_n)g=\lim_n xA(e_n)g=\lim_n gE(e_n)x=gEx=xE^*g.$$

Hence $\bigcup_n e_n\varepsilon N$, so N is a σ-algebra.

For every $E\varepsilon B$ define $f_E=T^{-1}E$. Since T^{-1} is an algebra isomorphism, $f_E^2=f_E$. Thus f_E is the characteristic function of a set $h(E)\subset S$. Since $f_E\varepsilon C(S)$, the set $h(E)$ is open and closed in the Gelfand topology of S. Since the topology of S coincides with the topology induced by $C(S)$, the sets

$$H(s_o;D;\alpha)=\{s\varepsilon S;\,|(T^{-1}D)(s)-(T^{-1}D)(s_o)|<\alpha,\ s_o\varepsilon S,\ D\varepsilon U(B),\ \alpha>0\}$$

are a subbase for the topology of S. B generates $U(B)$ in the uniform operator topology, therefore, by [3; IX.2.12], the sets $H(s_o;D;\alpha)$ with $D\varepsilon B$ form also a subbase for the topology of S. This means that the sets $h(E)$ $(E\varepsilon B)$ form a subbase and, since $h(E)h(F)=$ $=h(EF)$, they form a base.

Making use of the facts that the complex-valued set function $xA(\cdot)x^*$ is regular for every $x\varepsilon X$, $x^*\varepsilon X^*$, and that $\{h(E);E\varepsilon B\}$ is a base for the topology of S, the following statement can be proved as in the proof of [3; XVII.3.9].

Let e be an open subset of S, let $x\varepsilon X$, $x^*\varepsilon X^*$ and $\varepsilon>0$. Then there is a projection E_o in B such that $h(E_o)\subset e$ and for any E

in B such that $h(E_o) \subset h(E) \subset e$, we have

$$|x^*Ex - xA(e)x^*| < \varepsilon.$$

Let $\{\varepsilon_n\}$ be a sequence of positive numbers converging to 0, let E_{on} be the corresponding projections occurring in the statement above, and let $F_n = \bigvee_{i=1}^{n} E_{oi}$. Then $\{F_n\}$ is an increasing sequence in B, and $h(E_{on}) \subset h(F_n) \subset e$, therefore

$$|x^*F_nx - xA(e)x^*| < \varepsilon_n \qquad (x\varepsilon X,\ x^*\varepsilon X^*,\ n=1,2,\ldots).$$

Let $F = \bigvee_{n=1}^{\infty} F_n$. Since B is G-σ-complete,

$$\lim_n gF_nx = gFx \qquad (x\varepsilon X,\ g\varepsilon Q).$$

Therefore $(Fx,g) = (x,A(e)g)$ for $x\varepsilon X$, $g\varepsilon Q$, hence $F' = A(e)$. Thus $F = E(e)$ and $e\varepsilon N$. Since the σ-algebra N contains every open set e in M, we have N=M.

Now we show that the set function $E(\cdot)$ constructed above has the stated properties. The duality (2) shows that E is a spectral measure of class (M,G). For any $x\varepsilon X$, $g\varepsilon Q$

$$(T(f)x,g) = \int_S f(s)(x,\ A(ds)g) = \int_S f(s)(x, E(ds)^*g) = \left(\int_S f(s)E(ds)x, g\right)$$

proves that $T(f) = \int_S f(s)E(ds)$. The duality (2) shows again that $gE(\cdot)x = xA(\cdot)g$ is a regular set function for any $x\varepsilon X$, $g\varepsilon G$. If $D(\cdot)$ is another representing measure with the stated properties, then for every $f\varepsilon C(S)$, $x\varepsilon X$, $g\varepsilon G$

$$\int_S f(s)((E(ds) - D(ds))x,g) = 0.$$

By regularity and the Riesz representation theorem

$$(E(e)x,g) = (D(e)x,g) \qquad (e\varepsilon M),$$

which proves uniqueness. By construction, the range $E(M)$ is contained in B. Conversely, if $P\varepsilon B$ then

$$P = T(f_P) = \int_S f_P(s)E(ds) = E(h(P)).$$

The proof is complete.

THEOREM 2. *A Boolean algebra of projections in* X *is* G-σ-*complete if and only if it coincides with the range of a spectral measure of class* (M,G) *for some* σ-*algebra* M *of subsets of a com-*

pact Hausdorff space.

PROOF. Lemmas 1 and 2.

For σ-complete Boolean algebras of projections a considerable part of the theory is based on the existence of certain positive functionals, called Bade functionals (cf.[3;XVII.3.12]). If B is a σ-complete Boolean algebra of projections in X, then for every x in X there is x^* in X^* (a Bade functional for x) such that $x^*Ex \geq 0$ for every $E \in B$, and $x^*E_o x=0$ for some E_o in B implies $E_o x=0$. However, if B is only G-σ-complete for some total linear manifold G in X^*, it can happen that no Bade functionals exist for a given x in X (see Section 3).

DEFINITION 3. Assume that B is G-σ-complete, and let $x \in X$, $x^* \in X^*$. Define $Q=B^*G$ and

$$Q(x)=Q\text{-sp}(Ex;\ E \in B), \quad X(x^*)=X\text{-sp}(E^*x^*;\ E \in B).$$

If $Q(x)=X$, then x will be called a G-*cyclic* (or Q-*cyclic*) *vector* for B. If $X(x^*)=X^*$, then x^* will be called an X-cyclic vector for B^*.

REMARK. If $G=X^*$, i.e. B is σ-complete in the sense of Bade and x is an (X^*-) cyclic vector for B then, by [3;XVII.3.13], any Bade functional x^* for x is an X-cyclic vector for B^*. However, in the general case ($G \neq X^*$) it can happen that there is an X^*-cyclic vector for B, but there is no X-cyclic vector for B^* (see Section 3).

DEFINITION 4. If B is a G-σ-complete Boolean algebra of projections and $g \in B^*G$, then define

$$K(g)=\{x \in X;\ gEx=0 \text{ for all } E \in B\}= \\ =\bigcap\{\ker (E^*g);\ E \in B\}.$$

Note that, since a linear manifold in X^* is total if and only if it is $w(X^*, X)$-dense, $K(g)=\{0\}$ if and only if g is an X-cyclic vector for B^*. Further, we always have $\bigcap\{K(g);\ g \in G\}=\{0\}$.

THEOREM 3. *Let* B *be a* G-σ-*complete Boolean algebra of projections in* X, *and let* $Q=B^*G$. *Assume that there is a* G-*cyclic vector* x *for* B *and that there exist* $(g_i, E_i) \in Q \times B$ $(i=1,2,\dots)$ *such that*

$$K(g_i) \subset E_i X \quad (i=1,2,\ldots) \text{ and } \bigwedge_i E_i = 0.$$

Then $U(B)=B^c$.

PROOF. Since $U(B) \subset B^c$, we have to prove the converse inclusion. By Theorem 2, B coincides with the range of a spectral measure E of class (M,G), where M is a σ-algebra of subsets of a set S. Let $A \varepsilon B^c$, $g \varepsilon \{g_i; i=1,2,\ldots\}$, $b(e)=gAE(e)x$ and $m(e)=gE(e)x$ $(e \varepsilon M)$. Because b and m are countably additive complex valued set functions on M, it follows that b is m-continuous if and only if $v(m,e)=0$ implies $b(e)=0$ ([3;III.4.13]), where v denotes the total variation of m.

If $v(m,e)=0$ then $gE(d)x=0$ for any $d \subset e$, $d \varepsilon M$. Since x is G-cyclic for B, for any x_o in X there are sets $f_i^a \varepsilon M$ such that

$$x_o = Q\text{-}\lim_a \sum_{i=1}^{n(a)} E(f_i^a)x.$$

Since $E(e) \varepsilon B$, it is $w(X,Q)$-$w(X,Q)$-continuous. Hence $E(e)x_o = Q$-$\lim_a \sum_i E(f_i^a e)x$, in particular

$$gE(e)x_o = \lim_a \sum_i gE(f_i^a e)x = 0.$$

Since $A \varepsilon B^c$, $AE(e)x = E(e)Ax \varepsilon E(e)X \subset \ker(g)$, therefore $b(e)=0$, and b is m-continuous. By the Radon-Nikodym theorem there is an M-measurable, m-integrable function $h=h(g;\cdot)$, defined m-almost everywhere and such that

$$gAE(e)x = \int_e h(g;s)gE(ds)x \qquad (e \varepsilon M).$$

Pick one representative of h (which will be denoted in the same manner) and define

$$e_n(g) = \{s \varepsilon S; |h(g;s)| \le n\}, \qquad A_n(g) = \int_{e_n(g)} h(g;s)E(ds).$$

Then for any $e,d \varepsilon M$

$$gE(e)(E(e_n)A - A_n)E(d)x = \int_{ee_n d} (h(s)-h(s))gE(ds)x = 0.$$

Since A, $A_n \varepsilon B^c$, therefore $y^* = (A^*E(e_n)^* - A_n^*)E(e)^* g \varepsilon Q$. Hence y^* is $w(X,Q)$-continuous. Since x is a Q-cyclic vector for B, $y^*=0$. So

$$E(e)^*g\ (E(e_n)A - A_n) = 0 \qquad \text{for every } e \varepsilon M.$$

Hence there is $B_n(g) \varepsilon B(X)$ such that

$$E(e_n(g))A = A_n(g) + B_n(g) \qquad \text{and} \qquad B_n(g)X \subset K(g).$$

Since for some i we have $g=g_i$, by assumption $K(g) \subset E_i X$. Introducing the notation $F_i=I-E_i$, we have $F_i=E(d_i)=E(d_i(g_i))$, and

$$E(e_n(g_i)d_i)A=E(d_i)A_n(g_i)=\int_{e_n(g_i)d_i} h(g_i;s)E(ds).$$

Since the algebra B is bounded in the (uniform) topology of B(X), there is p>0 such that

$$|E(d_i)A_n(g_i)|\leq p \qquad \text{for every } i,n=1,2,\ldots .$$

From the proof of [3;XVII.2.10] it is seen that

$$E\text{-}\operatorname{ess\,sup}_{s\in e_n(g_i)d_i} |h(g_i;s)|\leq|E(d_i)A_n(g_i)|\leq p \qquad (i,n=1,2,\ldots).$$

Hence

$$E\text{-}\operatorname{ess\,sup}_{s\in d_i}|h(g_i;s)|\leq p \qquad (i=1,2,\ \ldots).$$

Therefore there exists $\int_{d_i} h(g_i;s)E(ds)$ $(i=1,2,\ \ldots)$, and

$$qE(e_n(g_i))E(d_i)Ax=qE(e_n(g_i))\int_{d_i} h(g_i;s)E(ds)x \qquad (q\varepsilon Q,\ x\varepsilon X).$$

Hence $E(d_i)A=\int_{d_i} h(g_i;s)E(ds)$. For every $i,k=1,2,\ \ldots$

$$\int_{d_id_k} h(g_i;s)E(ds)=AE(d_id_k)=\int_{d_kd_i} h(g_k;s)E(ds).$$

Since $E\text{-}\operatorname{ess\,sup}_{s\varepsilon S}|f(s)|\leq|\int_S f(s)E(ds)|$ for any $f\varepsilon EB(S,M)$, for $i,k=1,2,\ \ldots$ we have

$$h(g_i;s)=h(g_k;s) \qquad \text{E-almost everywhere on } d_id_k.$$

Let

$$h(s)=h(g_i;s) \qquad \text{for } s\varepsilon d_i,\quad i=1,2,\ \ldots .$$

Since $E(\bigcup_i d_i)=\bigvee_i F_i=I$, as we have seen above h is well-defined E-almost everywhere, and is E-essentially bounded by p. Define

$$\bar{A}=\int_S h(s)E(ds).$$

Let $c_k=\bigcup_{i=1}^{k} d_i$. Then

$$E(c_k)A=\int_{c_k} h(s)E(ds)=E(c_k)\bar{A} \qquad (k=1,2,\ \ldots).$$

Since $E(\bigcup_k c_k)=I$, therefore $A-\bar{A}=Q\text{-}\lim_k E(c_k)(A-\bar{A})=0$ and A belongs to U(B). The proof is complete.

COROLLARY. *Let* B,Q *be as above. Assume that there is a* G-*-cyclic vector* x *for* B *and that there is an* X*-cyclic vector* $q\varepsilon Q$

for B^*. *Then* $U(B)=B^c$.

3. AN EXAMPLE

Let $S=[0,1]$ and let M be the σ-algebra of the Borel subsets of S. Let $X=B(S,M)$ be the space of all uniform limits of finite linear combinations of characteristic functions of sets in M, i.e. the set of all bounded Borel-measurable functions on S, with norm

$$|f| = \sup_{s\in S} |f(s)|,$$

(cf. [3;IV.2.12]). By [3;IV.5.1], $B(S,M)^*=ba(S,M)$, the space of all bounded additive scalar functions defined on M, with norm $|m|=v(m,S)$, the total variation of m on S. For any $f\in X$ let

$$(E(b)f)(s)=k(b;s)f(s) \qquad (b\in M)$$

where $k(b;\cdot)$ denotes the characteristic function of the set b. Then $\{E(b);\ b\in M\}$ is a G-σ-complete Boolean algebra of projections in X, where $G=ca(S,M)$, the total linear manifold of countably additive set functions in X^*. Note that E is the resolution of the identity of class G of the prespectral operator T defined by $(Tf)(s)=sf(s)$.

Assume now that $x(s)\equiv 1$ for s in S and x^* is a Bade functional for x. Then

$$0\le x^*E(b)x=\int_b x^*(ds)=x^*(b) \qquad \text{for } b\in M,$$

and $x^*(b)=0$ implies $E(b)x=0$, i.e. that $b=\emptyset$. Hence $x^*(\{s\})>0$ for any $s\in[0,1]$, and $x^*\in ba\ (S,M)$, which is absurd. Therefore:

(a) There is no Bade functional for x.

Further, we clearly have

$$X=\overline{sp}\ (Ex;E\in B).$$

Assume that x_o^* is an X-cyclic vector for B^*. Then $(E(b)^*x_o^*)y_o=0$ for every b in M and for some y_o in X should imply $y_o=0$. Clearly, $v(x_o^*;\{s\})=0$ for several s in S, so there is $y\in X$, $y\neq 0$ such that y is an x_o^*-null function. Hence $(E(b)^*x_o^*)y=\int_b y(s)x_o^*(ds)=0$ for every $b\in M$. Therefore:

(b) There is a strongly ($=X^*$-) cyclic vector for B, but there is no X-cyclic vector in X^* for B^*.

By definition, the sets $K(g)$ consist of those x for which

$\int_b x(s)g(ds)=0$ $(b\varepsilon M)$, i.e. for which $x(s)=0$ g-almost everywhere. Since $v(g;\{s\})>0$ can hold only for points s in some countable set $H(g)$ in S, if $K(g)\subset E(b)X$ for some $b\varepsilon M$ then b has a countable complement in S. Therefore the asumption of Theorem 3 does not hold for any countable set $\{(g_i,E_i)\}$.

However, we still have $U(B)=B^c$. Indeed, if $T\varepsilon B^c$ and $f_1(s)\equiv 1$ for s in S, then $T(k(b))=k(b)Tf_1$ for any Borel set b. Hence $Tf=f(Tf_1)$ for any $f\varepsilon X$. With the notation $Tf_1=f_2$ we have $(Tf)(s)=$ $=f_2(s)f(s)$ $(f\varepsilon X)$.Therefore, if $e(s)$ is a finite linear combination of characteristic functions of sets in M and E is the operator defined by $(Ef)(s)=e(s)f(s)$, we have

$$|(T-E)f|\leq|f_2-e||f| \qquad \text{for every } f\varepsilon X.$$

Since $E\varepsilon U(B)$, and the family of finite linear combinations above is dense in X, we obtain that T belongs to $U(B)$. So

(c) The condition of Theorem 3 ensuring that $U(B)=B^c$ is sufficient but not necessary.

REMARK. If in the example above the basic space X is changed from $B(S,M)$ to $L^\infty(S,M,p)$ where p is Borel measure on M,then the vector $x_1\equiv 1$ is strongly cyclic for B, which is G-σ-complete with $G=L^1(S,M,p)$.For any x in X, the vector $\bar{x}$ in G, represented by the complex conjugate of x, is a Bade functional for x. Indeed,

$$\int_e \bar{x}(t)x(t)\ p(dt)\geq 0 \qquad (e\varepsilon M),$$

and if the above integral is 0 for some $e_o\varepsilon M$, then $E(e_o)x=0$.Further, the vector $x_1\varepsilon G$ is X-cyclic for B^*. Hence the Corollary to Theorem 3 applies, and $U(B)=B^c$.

4. ON PRESPECTRAL OPERATORS

Let M denote the σ-algebra of all Borel sets in the complex plane C and let G be a total linear manifold in the dual X^* of the complex Banach space X. Recall that an operator T, in $B(X)$, is called *prespectral of class* G if there is a (then unique) spectral measure E of class (M,G) (called the resolution of the identity of class G for T) such that for all Borel sets b

$$TE(b)=E(b)T \qquad \text{and} \qquad \sigma(T|E(b)X)\subset\bar{b}$$

(σ will denote the spectrum of the operator). For the fundamental properties of prespectral and spectral operators (the latter are the prespectral operators of class X^*) we refer to the monographies of Dowson [2] and Dunford and Schwartz [3].

If E is the resolution of the identity of class G for T, then the Boolean algebra $B=\{E(b);b\epsilon M\}$ is G-σ-complete, by Lemma 1. We shall denote the linear manifold B^*G by Q.

Let T be a prespectral operator on X with resolution of the identity E of class G. Define

$$S(G)=\int_{\sigma(T)} z\,E(dz), \qquad N(G)=T-S(G).$$

By [2; Theorem 5.13], the operators $S(G)$, hence $N(G)$, are independent of the class G. The operator $S=S(G)$ is called the *scalar part* and the operator $N=N(G)$ is called the *radical part* of T. The decomposition

$$T=S+N$$

is said to be the *Jordan decomposition* of T. By [2; Theorem 5.22], the adjoint T^* is then prespectral on X^* with resolution of the identity F of class X such that with the notation

$$h(f)=\int_{\sigma(T)} f(z)E(dz) \qquad (f\epsilon C(\sigma(T)))$$

we have

$$h(f)^*=\int_{\sigma(T)} f(z)F(dz) \qquad (f\epsilon C(\sigma(T))).$$

Moreover,

$$T^*=S^*+N^*$$

is then the Jordan decomposition of the operator T^*.

LEMMA 3. *Let* $T\epsilon B(X)$ *be a prespectral operator with radical part* N *and with resolution of the identity* E *of class* G, *and let* F *denote the resolution of the identity of class* X *for* T^*. *Then*

$$F(b)g=E(b)^*g \qquad \textit{for all } b\epsilon M, \quad g\epsilon Q.$$

Further, $F(b)Q\subset Q$ *for every* b *in* M, *and* $N^*Q\subset Q$.

PROOF. Let K denote $\sigma(T)$, let $c\subset K$ be a compact set, let $d(z,c)$ denote the distance of the point z from c, and let n be a positive integer. Define

$$f_n(z)=1-\min(1-n^{-1}, nd(z,c)) \qquad (z\epsilon K).$$

Then $f_n\epsilon C(K)$, and

$$|f_n(z)|\le 1, \ \lim_n f_n(z)=k(c;z) \qquad (z\epsilon K),$$

where $k(c;z)$ is the value of the characteristic function of c at z. Let $x\in X$, $g\in Q$, $m(b)=(x,F(b)g)$ and $p(b)=(E(b)x,g)$ for $b\in M$. Then m and p are countably additive, therefore Lebesgue's dominated convergence theorem yields

$$(E(c)x,g)=\int_K k(c;z)p(dz)=\lim_n\int_K f_n(z)p(dz)=\lim_n (h(f_n)x,g)=$$
$$=\lim_n (x,h(f_n)^*g)=\lim_n\int_K f_n(z)m(dz)=\int_K k(c;z)m(dz)=(x,F(c)g).$$

Hence $F(c)g=E(c)^*g$ for every compact c. Let

$$D=\{b\in M;\ (E(b)x,g)=(x,F(b)g) \quad \text{for every } x\in X,\ g\in Q\}.$$

We have that D is a σ-algebra containing the compact subsets. Therefore

$$F(b)g=E(b)^*g \qquad \text{for every } b\in M,\ g\in Q.$$

The last assertion of the lemma follows from this and from the facts that $E(b)\in B^c$ for every b in M and $N\in B^c$.

COROLLARY. *Assume that* $T\in B(X)$ *is prespectral with resolution of the identity* E *of class* G, *and let* E_2 *denote the resolution of the identity of class* X^* *for* T^{**}. *Then for every* $g\in Q$, $x\in X$, $b\in M$

$$(g,\ E(b)x)=(g,\ E_2(b)x).$$

PROOF. Let E_1 denote the resolution of the identity of class X for T^*. Then, by Lemma 3, $(g,E(b)x)=(E_1(b)g,x)=(g,E_2(b)x)$.

The following theorem will use the notations of Lemma 3. The lemma shows that the assumptions that G is F-invariant and that $N_1G\subset G$ are not essential restrictions.

THEOREM 4. *Let* $T\in B(X)$. *Assume that* T^*, *in* $B(X^*)$, *is prespectral with radical part* N_1, *with resolution of the identity* F *of class* X *and that* G *is a total* F*-invariant linear manifold in* X^* *such that* $N_1G\subset G$. T *is a prespectral operator of class* G *if and only if for every Borel set* b

$$F(b)|G\in C(w(G,X),\ w(G,X)),$$

further

$$N_1\in C(w(X^*,X),\ w(X^*,X)).$$

PROOF. Assume first that T is prespectral with resolution of the identity E of class G. Let the linear spaces X and G be given their natural pairing: $(x,g)=g(x)$ for $x\in X$ and $g\in G$. By Lemma 3 and by assumption, $E(b)^*g=F(b)g\in G$ for every $b\in M$, $g\in G$. Then for each Borel set b the transformations $E(b):X\to X$ and $F(b)|G:G\to G$ are duals with respect to the pairing above. By [4; 21.1], each dual $F(b)|G$ is $w(G,X)$-$w(G,X)$-continuous. By [2; Theorem 5.22], N_1 is the adjoint N^* of the radical part N of T, hence $N_1\in C(w(X^*,X), w(X^*,X))$.

Conversely, assume now that each restriction $F(b)|G$ is $w(G,X)$-$w(G,X)$-continuous. Since G is total in X^*, G distinguishes points in X with respect to their natural pairing. By [4;21.1], the dual of $F(b)|G$ with respect to this pairing of G and X is a uniquely determined $w(X,G)$-$w(X,G)$-continuous linear transformation, which will be denoted by $E(b)$ $(b\in M)$. According to this duality,

(*) $\qquad (F(b)g,x)=(g,E(b)x) \qquad$ for all $b\in M$, $g\in G$, $x\in X$.

Hence $gE(b)=F(b)g\in G$, so $gE(b)$ is continuous as a mapping of the Banach space X to C. Since G is total, [3; II.2.7] yields that $E(b)\in B(X)$.

It is easily seen that $E(\cdot)$ satisfies the conditions (i)-(iii) of a spectral measure of class (M,G). For example,

$$(g,E(bc)x)=(F(bc)g,x)=(F(b)F(c)g,x)=(g,E(b)E(c)x)$$

and the fact that G is total, proves (iii). From (*) we also obtain (iv), since $F(\cdot)$ is countably additive in the X operator topology. Therefore E is a spectral measure of class (M,G).

By assumption, N_1 is the adjoint N^* of an operator $N\in B(X)$. Since N_1 is quasinilpotent, so is N. Let $T^*=S_1+N_1$ be the Jordan decomposition of the prespectral operator T^*. Then $S_1=T^*-N^*$ is the adjoint A^* of the operator $A=T-N\in B(X)$. Define

$$S=\int z\ E(dz)=\int_{\sigma(T)} z\ E(dz).$$

For any $x\in X$, $g\in G$ then

$$(Sx,g)=\int z(E(dz)x,g)=\int z(x,F(dz)g)=(x,S_1g)=(Ax,g).$$

Since G is total, S=A. By assumption, $N^*G\subset G$, therefore

$$(E(b)Nx,g)=(Nx,F(b)g)=(x,N^*F(b)g)=(x,F(b)N^*g)=$$
$$=(NE(b)x,g) \qquad (x\in X,\ g\in G,\ b\in M).$$

Hence N commutes with each E(b). The operator T is the sum of the scalar operator S with resolution of the identity E of class

G and of the quasinilpotent N commuting with each E(b). By [2; Theorem 5.15], T is prespectral of class G. Then proof is complete.

COROLLARY. *Assume that* $T\varepsilon B(X)$ *is prespectral of classes* G_1 *and* G_2 *such that the resolution of the identity* F *of class* X *for* T^* *and the radical part* N^* *of* T^* *leave the total linear manifolds* G_1, G_2 *invariant. Let* $G=G_1+G_2$ *and assume that for each net* $\{g_a\}$ *in* G *converging to* $g\varepsilon G$ *in the* $w(G,X)$ *topology there are two nets* $\{g_a^i\}$ *in* G_i *converging to* $g^i\varepsilon G_i$ $(i=1,2)$ *in the same topology and such that* $g_a=g_a^1+g_a^2$ *for every* a. *Then* T *is prespectral of class* G.

THEOREM 5. *Assume that the adjoint* T^* *of the operator* $T\varepsilon B(X)$ *is prespectral with resolution of the identity* E_1 *of class* X. *Then the following statements are equivalent:*

(1) T *is a spectral operator;*

(2) *For each Borel set* b

$$E_1(b)\varepsilon C(w(X^*,X),\ w(X^*,X));$$

(3) *The resolution of the identity* E_2 *of class* X^* *for the operator* $T^{**}\varepsilon B(X^{**})$ *satisfies for each Borel set* b *the relation* $E_2(b)X\subset X$.

PROOF. If T is spectral with resolution of the identity E, then $E_1(b)=E(b)^*$ for each Borel set b. Hence (1) implies (2). If (2) holds, then each $E_1(b)$ is the adjoint of some operator $E(b)\varepsilon B(X)$. By the proof of Theorem 4, E is a spectral measure of class (M,X^*), where M denotes the σ-algebra of Borel sets. Let $T^*=S_1+N_1$ be the Jordan decomposition of T^* and let

$$S=\int z\ E(dz)=\int_{\sigma(T)} z\ E(dz).$$

Then

$$S_1=\int z\ E(dz)^*=(\int z\ E(dz))^*=S^*,$$

hence $N_1=T^*-S^*$ is the adjoint N^* of the operator $N=T-S$. Since N_1 commutes with S_1, the quasinilpotent operator N commutes with S, which is a spectral operator of scalar type. Therefore T is a spectral operator with resolution of the identity E, so (1) and (2) are equivalent.

If T is a spectral operator, then Corollary to Lemma 3 shows that

$$E_2(b)x=E(b)x \qquad (b\varepsilon M,\ x\varepsilon X),$$

so (1) implies (3).

If (3) holds, then the restriction E to X of E_2 is a spectral measure of class (M,X^*) on X. Let $T^{**}=S_2+N_2$ be the Jordan decomposition of the prespectral operator T^{**}, then

$$S_2=\int z\ E_2(dz)=\int_{\sigma(T)} z\ E_2(dz).$$

The restriction S of S_2 to X satisfies

$$S=\int z\ E(dz),$$

so S is a spectral operator of scalar type. The operator $N=N_2|X=$ $=T-S$ is quasinilpotent and, since N_2 and S_2 commute, N commutes with S. Hence T is a spectral operator in B(X) with resolution of the identity E, thus (1) and (3) are equivalent. The proof is complete.

REFERENCES

1. Bade,W.G.: On Boolean algebras of projections and algebras of operators, *Trans.Amer.Math.Soc.* 80(1955), 345-360.

2. Dowson,H.R.: *Spectral theory of linear operators*, Academic Press, London, 1978.

3. Dunford,N.; Schwartz,J.T.: *Linear operators*, Part I: 1958, Part II: 1963, Part III: 1971, Wiley-Interscience, New York.

4. Kelley,J.L.; Namioka, I.; et al.: *Linear topological spaces*, Van Nostrand, Princeton, 1963.

5. Palmer,T.W.: Unbounded normal operators on Banach spaces, *Trans.Amer.Math.Soc.* 133(1968), 385-414.

B.Nagy
Department of Mathematics,
Faculty of Chemical Engineering,
University of Technology,
Budapest, Hungary.

CONTROL SUBSPACES OF MINIMAL DIMENSION, AND SPECTRAL MULTIPLICITIES

N.K.Nikol'skii and V.I.Vasjunin

1. Introduction.
2. Preliminaries and definitions.
3. Elementary examples (shift, backward shifts, simple unitary operators).
4. Isometries, coisometries, unitary operators, C_0-contractions.
5. Normal operators.
6. Complete, compact, strictly upper triangular operators; Toeplitz operators.
7. Some unsolved problems.

1. INTRODUCTION

Let us consider a linear dynamical system

$$(*) \qquad \dot{x}(t)=Ax(t)+Bu(t), \qquad t\geq 0,$$

where $A:X\to X$, $B:U\to X$ are bounded linear operators, X (the state space) and U (the control or input space) are some normed linear spaces.

The general problem of control is to describe such operator pairs (A,B) for which there exists an appropriate input signal $u(\cdot)$ such that the system comes in a moment t into prescribed neighbourhood of an arbitrary "state" x, $x\in X$, from a fixed initial state $x(0)$ (say $x(0)=0$). After specifying the terminology this problem is easily solvable (R.E.Kalman): the system is controllable iff the subspace BU is cyclic for A. If the system has the last property we can raise a further problem, namely to minimize the dimension of the control subspace without loss of controllability. That is we have to find a subspace $U'\subset U$ -the smaller the better - but such that the system with the pair $(A,B|U')$ is still controllable. The desired smallness of U' can be measured by the dimension $\dim BU'$. If for a given U we find $\min \dim BU'$ and then we maximize this quantity over all subspaces BU, $BU \subset X$, cyclic for A, we obtain a number depending on operator A only, but not on B or U. This number we denote by disc A (dimension of the input subspace of control).

Principal aims of this report are a description of basic properties of the functional disc A and its computation for some useful classes of operators (unitary operators, isometries and co-isometries, Sz.-Nagy-Foiaş model operators, Toeplitz operators, compact operators and some others). We try, in particular, to understand the connection of disc A with the usual spectral multiplicity μ_A and the properties of the lattice Lat A of invariant subspaces influencing disc A.

Some theorems are only stated here; their proofs can be found in [1],[2],[3]. We end with a list of unsolved problems.

The authors are grateful to V.P.Havin for his numerous language suggestions.

2. PRELIMINARIES AND DEFINITIONS

Let us begin with the definition of controllability.

2.1. DEFINITION. The system (*) is *controllable* if for any $x \in X$ and any $\varepsilon > 0$ there exist $t > 0$ and a smooth function $u:[0,\infty) \to U$ such that the equation (*) has a solution $x(\cdot)$ with the following properties:

$$x(0)=0, \qquad \|x(t)-x\|<\varepsilon.$$

It is clear that in the case $\dim BU < \infty$ (only this physically interesting case will be considered here) the controllability of the system depends on the subspace BU only, but not on B or U.

An analogous definition of the controllability can be stated for the discrete system

$$(**) \qquad x_{n+1} = Ax_n + Bu_n, \qquad n \geq 0.$$

It is worth noting that in the classical finite dimensional case ($\dim X < \infty$) the linearity of the set of reachable states (i.e. of the set $\{x(t): \forall t>0,\ \forall u(\cdot)\}$) implies that the system is controllable (i.e. $\forall x \in X\ \exists t>0\ \exists u(\cdot): x(t)=x$).

The following theorem reduces the controllability problem to the investigation of cyclic subspaces of A.

2.2. THEOREM. *Let* X,U *be Banach spaces;* $A:X\to X$, $B:U\to X$ *be bounded linear transforms. The following assertions are equivalent:*

1) *The system* (*) *is controllable.*
2) *The system* (**) *is controllable.*
3) Span $\{A^n BU: n\geq 0\} = X$.

This theorem is essentially due to R.E.Kalman.

Let us note that the boundedness of A is not necessary here; it may be replaced by the strong continuity of the semigroup $e^{sA}(s\geq 0)$. Then the condition 3) is to rewrite in another form: instead of powers A^n we have to use the resolvent of A or the semigroup e^{sA} itself.

2.3. DEFINITION OF disc A. Let us denote

$$E_R = E_R^A \stackrel{\text{def}}{=\!=} \text{span}\{A^n R: n\geq 0\}, \qquad \text{where } R\subset X.$$

The system is controllable if $E_R = X$ for $R=BU$. For such system, we can try to choose a subspace R' of R with the property $E_{R'} = X$. The number

$$\min\{\dim R' : R'\subset R,\ E_{R'} = X\}$$

characterizes the most economical way of control inside of the given controllable system. And the quantity

$$\text{disc } A \stackrel{\text{def}}{=\!=} \sup[\min\{\dim R' : R'\subset R, E_{R'} = X\} : \dim R<\infty, E_R = X]$$

is an "index of the quality" of the transfer operator A. The condition $\dim R<\infty$ occurs here by two reasons. Firstly, only finite dimensional subspaces are of any interest in the control theory. Secondly, if this condition is dropped our supremum becomes infinite practically for any operator A. For an operator A without finite dimensional cyclic subspaces we assume disc $A=\infty$.

2.4. Disc AND THE SPECTRAL MULTIPLICITY. Disc A gives us an extra information about the operator A (not connected with the systems (*) and (**)). In particular the inequality disc $A<\infty$ presupposes, of course, the existence of finite dimensional cyclic subspaces of A, i.e. of subspaces R with the properties:

$$\text{(Cyc)} \qquad E_R^A = X, \qquad \dim R<\infty.$$

We denote the family of all such subspaces by Cyc A. In these terms disc $A=\sup[\min\{\dim R':R'\subset R,\ R'\in \mathrm{Cyc}\,A\}:R\in \mathrm{Cyc}\,A]$.
Recall that the spectral multiplicity of A is defined by the equality[1)]

$$\mu_A = \min\{\dim R: R\in \mathrm{Cyc}\,A\}.$$

It is worth noting that both disc A and μ_A depend on the lattice of invariant subspaces (Lat A) rather than on the operator A itself. To check this it is enough to note that

$$E_R^A = \bigcap_{\substack{E\supset R\\ E\in \mathrm{Lat}\,A}} E$$

for any $R\subset X$. Hence the equality Lat A=Lat B implies disc A = =disc B.

It is clear that

$$\mu_A \le \mathrm{disc}\,A.$$

It is not difficult to give example when μ_A=disc A (e.g. an arbitrary operator in a finite dimensional space) or - on the contrary - of the strong inequality $\mu_A<$disc A (e.g. for a unitary operator, if its absolutely continuous spectrum fills the whole unit disc, for an arbitrary nonunitary isometry of finite multiplicity and in an infinite dimensional space, of course). Speaking somewhat "qualitatively", one can say that the first of the mentioned possibilities occurs for operators A with many μ_A-dimensional cyclic subspaces, and the second one occurs for operators with poor supply of such subspaces.

Now we enlist some simple properties of μ_A and disc A.

2.5. If operators A and B are similar then $\mu_A=\mu_B$ and disc A= =disc B.

2.6. $\mu_A\ge \dim \mathrm{Ker}\,A^*$, $\mu_A \ge \sup\limits_{\lambda\in\mathbb{C}} \dim \mathrm{Ker}(A^*-\lambda I)$.

2.7. disc $A\oplus B\ge\max(\mathrm{disc}\,A,\ \mathrm{disc}\,B)$.

2.8. $E\in \mathrm{Lat}\,A$, $\mu_{A|E}<\infty \Rightarrow$ disc $A/E \le$ disc A, A/E being "the factor-operator " from X/E into X/E $((A/E)(x+E) \overset{\mathrm{def}}{=\!=} Ax+E,\ x\in X)$.

2.9. disc $A=\sup\limits_{R\in \mathrm{Cyc}\,A} \min\{\dim R'+\mathrm{disc}\,A/E_{R'} : R'\subset R\}$.

---

1) $\min\emptyset \overset{\mathrm{def}}{=\!=} +\infty$.

In particular

$$\operatorname{disc} A \le n + \sup_{R \in \operatorname{Cyc} A} \min\{\operatorname{disc} A/E_{R'} \,:\dim R'=n, R'\subset R\}.$$

2.10. OPERATORS WITH MANY CYCLIC SUBSPACES. Let $G_n(X)$ be the Grassmann manifold of all n-dimensional subspaces of X. For a finite dimensional space X there exists a normed measure τ_n invariant under unitary transforms of X. We shall say that the operator A has many cyclic subspace if for any $R \in \operatorname{Cyc} A$ almost all (with respect to τ_n, $n=\operatorname{disc} A$) subspaces $R'\subset R, \dim R'=n$, are cyclic.

Let $A=A_1 \oplus \dots \oplus A_n$ be an operator acting in the direct sum $X = X_1 \oplus \dots \oplus X_n$ of subspaces X_i, $A_i \in L(X_i)$, Pr_{X_i} be the natural projection X on X_i, $1\le i\le n$. It is clear that $R\in \operatorname{Cyc} A$ implies $Pr_{X_i} R \in \operatorname{Cyc} A_i$. If the converse is true we shall write

$$\operatorname{Cyc} A = \bigvee_{i=1}^{n} \operatorname{Cyc} A_i. \tag{1}$$

Obviously, (1) holds if

$$\operatorname{Lat} A = \sum_{i=1}^{n} \oplus \operatorname{Lat} A_i \tag{2}$$

i.e. every invariant subspace of A is the direct sum of invariant subspaces of A_i.

Let us note one more property of A that implies (2) and, consequently, (1). Denote by $\mathcal{R}_A$ the minimal weakly closed algebra containing A and I. Then the property is

$$\mathcal{R}_{\Sigma \oplus A_i} = \Sigma \oplus \mathcal{R}_{A_i} \tag{3}$$

i.e. $B_i \in \mathcal{R}_{A_i}$ implies the existence of $B \in \mathcal{R}_A$ such that $B = \sum_{i=1}^{n} \oplus B_i$.

This property can be interpreted as an assertion on the simultaneous approximation: it means the existence of a family of polynomials $\{p_\alpha\}$ with $\text{w-}\lim_\alpha p_\alpha(A_i)=B_i$, $\forall i$. Such a family exists if, for example, the polynomial hulls of spectra $\sigma(A_i)$ do not intersect, i.e. if there exist simply connected closed sets F_i such that

$$\sigma(A_i)\subset F_i,\ 1\le i\le n;\quad F_i \cap F_j = \emptyset \qquad \text{if } i\ne j. \tag{4}$$

For these operators $A=\Sigma \oplus A_i$ the conditions (1)-(3) hold.

Now we can state the following proposition.

2.11. *Let* $A = \sum_{i=1}^{n} \oplus A_i$. *If every operator* A_i *has many cyclic subspaces, and if* (1) *holds, then*

$$\text{disc } A = \max_{1\le i\le n} \text{disc } A_i$$

and A *has many cyclic subspaces too.*

Omitting the proof, we note only that the condition (1) is essential. For example, let A_o be an operator with the simple spectrum ($\mu_{A_o}=1$) in a finite dimensional subspace and $A_i=A_o$, $1\le i\le n$. Then disc $(\sum_{i=1}^{n} \oplus A_i)=n$, although A_o have many cyclic subspaces. It is not clear whether the second condition is essential. Dropping it we can prove only the following proposition.

2.12. *Let* $A = \sum_{i=1}^{n} \oplus A_i$. *Then*

$$\text{Cyc } A=\bigvee_{i=1}^{n} \text{Cyc } A_i \Rightarrow \text{disc } A \le \sum_{i=1}^{n} \text{disc } A_i.$$

Let us note that for the equality

$$\mu_A = \max_{1\le i\le n} \mu_{A_i}$$

the condition (1) is quite sufficient, and the inequality

$$\mu_A \le \sum_{i=1}^{n} \mu_{A_i}$$

is valid without any assumption.

2.13. FINITE DIMENSIONAL CASE. The following proposition is an immediate corollary of 2.11.

Let $A\in L(X)$, $\dim X<\infty$, *then*

$$\text{disc } A=\mu_A=\max_{\lambda\in\mathbb{C}} \dim \text{Ker } (A-\lambda I).$$

Moreover, A *has many cyclic subspaces.*

This formula was found by M. Heymann [6] using the theory of characteristic polynomials. We deduce it from 2.11. Indeed, using 2.5 we can assume that A is of the Jordan form, i.e.

$$A = \sum_{\lambda\in\sigma(A)} \oplus A_\lambda$$

where A_λ are Jordan cells, i.e. are cyclic operators with one point spectrum $\sigma(A_\lambda)=\{\lambda\}$. Therefore (4) and hence (1) are valid. Evidently A_λ have many ciclic subspaces and we can refer to 2.11.■

2.14. *If* $A\in L(X)$, *if* $\mu_A<\infty$ *and if the following implication*

$$E_i\in \mathrm{Lat}\, A,\ E_i\neq X,\ i=1,2 \Longrightarrow \mathrm{span}\{E_1,E_2\}\neq X$$

is valid then $\mathrm{disc}\, A=\mu_A=1$ *and almost all vectors are cyclic for* A.

We omit the easy proof of this proposition and state the following corollary.

2.15. *Let* A *be a unicellular operator* [1]*, then* $\mathrm{disc}\, A=\mu_A=1$ *and almost every vector is cyclic for* A.

The equality disc A=1 for a unicellular operator was noticed by Feintuch [5].

3. ELEMENTARY EXAMPLES

In this section we calculate disc of the following operators: the shift operator S in the Hardy space H^2, the backward shift S^* and its powers $S_n^*=S^*\oplus\ldots\oplus S^*$, a unitary operator without multiplicity and some Toeplitz operators.

3.1. OPERATOR S. Let $\mathbb{D}$ be the open unit disc of the complex plane, $\mathbb{T}$ be the unit circle, m be the Lebesgue measure on $\mathbb{T}$, $m(\mathbb{T})=1$. H^2 stands for the Hardy space of analytic functions f in $\mathbb{D}$ such that

$$||f||^2 \overset{\mathrm{def}}{=\!=} \sup_{0\le r<1}\int_{\mathbb{T}} |f(r\xi)|^2 dm(\xi)<\infty.$$

We consider H^2 as a subspace of $L^2(\mathbb{T})$, i.e. H^2 coincides with the subspace of functions whose Fourier coefficients with negative indices are zero. We define the operator S by the equality

$$(Sf)(\xi)=\xi f(\xi),\qquad f\in H^2,\qquad \xi\in\mathbb{D}.$$

Every isometric operator without unitary part and with simple spectrum is unitarily equivalent to S. By the Beurling theorem Lat S consists of all subspaces of the form

$$\Theta H^2=\{\Theta f\ :\ f\in H^2\}$$

where Θ is an inner function (i.e. $\Theta\in H^2$, and $|\Theta(\xi)|=1$ a.e. $\xi\in\mathbb{T}$).

1) i.e. $\forall E_1,E_2\in \mathrm{Lat}\, A$ either $E_1\subset E_2$ or $E_2\subset E_1$.

3.2. PROPOSITION. disc S=2 *and* S *has many cyclic subspaces.*

PROOF. Let us verify at first that disc S≥2. Let $0<|\lambda|<1$, $b_\lambda=(\lambda-z)(1-\bar{\lambda}z)^{-1}$ and $R=\mathrm{span}\,(z^n,b_\lambda)$, where n is chosen so that

$$\left(\frac{1+|\lambda|}{2}\right)^n < \min_{|\xi|=\frac{1+|\lambda|}{2}} |b_\lambda(\xi)| = \frac{1}{2+|\lambda|} .$$

It is clear that $R\in \mathrm{Cyc}\,S$ since $z^n\wedge b_\lambda=1$ [1]. But there is no cyclic function in R. Indeed, let $f=\alpha z^n+\beta b_\lambda$ and $|\alpha|\neq|\beta|$. Then f has a zero in $\mathbb{D}$ by Rouché's theorem (applied to the circle $\mathbb{T}$). If $|\alpha|=|\beta|$ we can apply Rouché's theorem to the circle $|z|=\frac{1}{2}(1+|\lambda|)$. In any case f has a zero in $\mathbb{D}$, therefore f is not outer, i.e. not cyclic. Hence disc S≥2.

The rest of the proof is based on the following criterion of cyclicity:

$$E_R^S=H^2 \Longleftrightarrow \wedge\{f^i: f\in R\}=1,$$

where f^i is the inner part of f (see [4]). We shall check (by induction with respect to dim R) that almost every two-dimensional subspace of an arbitrary finite dimensional cyclic subspace R is cyclic also.

<u>dim R=3.</u> Let f_1,f_2,f_3 be linearly independent functions from R. It is sufficient to prove that

$$f_1^i\wedge(f_2+\alpha f_3)^i=1 \qquad \text{for a.e. } \alpha\in\mathbb{C}.$$

Let $\delta_\alpha=f_1^i\wedge(f_2+\alpha f_3)^i$. Since $\delta_{\alpha_1}\wedge\delta_{\alpha_2}=f_1^i\wedge f_2^i\wedge f_3^i$ for $\alpha_1\neq\alpha_2$ and $f_1^i\wedge f_2^i\wedge f_3^i=1$, $\{\delta_\alpha:\alpha\in\mathbb{C}\}$ is a family of mutually disjoint divisors of f_1^i. Hence the set $\{\alpha:\delta_\alpha\neq 1\}$ is at most countable. Q.E.D.

<u>dim R>3.</u> Let $f\in R$ and let R′ be a complementing subspace, i.e. $f\cdot\mathbb{C}+R'=R$ and dim R′=dim R-1. Let $\theta=\wedge\{g^i:g\in R'\}$ and $R''=\theta^{-1}R'$. By the inductive assumption almost all two dimensional subspaces of R″ are cyclic. Therefore almost all two dimensional L's, $L\subset R'$, satisfy

$$\wedge\{g^i:g\in L\}=\wedge\{g^i:g\in R'\} .$$

Taking one of such L's, let us apply the preceding part of the

[1] The symbols ∧ and ∨ are used for the greatest common divisor and the least common multiple of a family of inner functions (for definitions see [4]).

argument to the three dimensional cyclic subspace $L=\text{span}(f,L)$. We see that almost all elements of $G_2(L)$ are cyclic, hence the same is true for almost all elements of $G_2(R)$.

3.3. OPERATORS S_n^*. It is adjoint to $S=S\oplus\ldots\oplus S$ (n times). We shall treat the orthogonal sum $H^2\oplus\ldots\oplus H^2=H_n^2$ as $H^2(\mathbb{C}^n)$, i.e. as the Hardy space of vector ($\mathbb{C}^n$-valued) functions. Then

$$S_n^*f=\frac{f-f(0)}{z},\qquad f\in H_n^2.$$

Remind that a function $f\in H^2$ is cyclic for S^* iff f has a meromorphic pseudocontinuation in $\hat{\mathbb{C}}\setminus\bar{\mathbb{D}}$, i.e. iff there exist bounded functions ϕ and ψ analytic in $\hat{\mathbb{C}}\setminus\bar{\mathbb{D}}$ and such that $f|\mathbb{T}=\frac{\phi}{\psi}|\mathbb{T}$ (see e.g. [4]). It is clear that no finite dimensional subspace of such functions is cyclic for S^*. The same is true for the operator S_n^* also, if we define pseudocontinuation for vector functions analogously: there exist a bounded vector function ϕ, analytic in $\hat{\mathbb{C}}\setminus\bar{\mathbb{D}}$, and a scalar one ψ so that $f|\mathbb{T}=\frac{\phi}{\psi}|\mathbb{T}$.

Now we prove the following statement.

3.4. PROPOSITION. disc $S_n^*=n$, *and* S_n^* *has many cyclic subspaces* [1].

PROOF. At first we check that disc $S_n^*\geq n$. Let $e_1,\ldots,e_n$ be the natural basis of $\mathbb{C}^n$, f be a cyclic function for S (e.g. $f=\log(1+z)$) and $R=f\cdot\mathbb{C}^n=\text{span}\,(fe_i:1\leq i\leq n)$. Clearly $R\in\text{Cyc}\,S_n^*$ and for every $R'\subset R$ with $\dim R'<n$, there exists a proper subspace $L\subset\mathbb{C}^n$ such that $R'=f\cdot L$. Hence

$$E_{R'}=\text{span}\{S_n^{*k}R':k\geq 0\}=H^2(L)\subsetneq H_n^2,$$

i.e. $R'\notin\text{Cyc}\,S_n^*$. Therefore disc $S_n^*\geq n$.

The converse inequality for n=1 follows from 2.14. Suppose n>1 and the proposition is true for every smaller value of n. If $R\in\text{Cyc}\,S_n^*$ then there exists $f\in R$ with no pseudocontinuation. Take such f and verify that disc $(S_n^*/E_f)\leq n-1$. Then by 2.9 we obtain disc $S_n^*\leq 1+(n-1)=n$.

By Lax-Halmos theorem (see e.g. [4]) the invariance of $E_f^\perp$

[1] Recall that $\mu_{S_n^*}=1$, $\forall n$.

under S_n implies $E_f^{\perp}=\Theta H_m^2$ for a m, m≤n and an inner function Θ [1]. Furthermore m<n, because $f\in H_n^2 \ominus \Theta H_m^2$ and f does not admit a pseudocontinuation. It is clear that S_n^*/E_f is unitarily equivalent to $P_{E_f^{\perp}} S_n^* | E_f^{\perp}$, which is in his turn unitarily equivalent to S_m^* and so its disc is equal to m, m≤n-1.

An analysis of this construction shows that almost all n dimensional subspaces of a given cyclic subspace are cyclic too. ∎

3.5. UNITARY OPERATORS WITHOUT MULTIPLICITY. By the Spectral Theorem every such operator U is unitarily equivalent to the following one:

$$f \to zf, \qquad f \in L^2(\mathbb{T},\nu),$$

where ν is the scalar spectral measure corresponding to U. Let $\nu_a+\nu_s$ be the decomposition of ν into the sum of its absolutely continuous and singular (with respect to the Lebesgue measure) parts. Correspondingly, we have

$$L^2(\mathbb{T},\nu)=L^2(\mathbb{T},\nu_a)\oplus L^2(\mathbb{T},\nu_s),$$
$$U=U_a\oplus U_s,$$
$$\mathrm{Lat}\ U=\mathrm{Lat}\ U_a\oplus \mathrm{Lat}\ U_s.$$

We remind that an operator U is called reductive, if Lat U⊂Lat U*. For a unitary operator this inclusion holds iff the measure m is not absolutely continuous with respect to the spectral measure of the operator (J.Wermer [8]; see also [4]).

3.6. PROPOSITION. *Let* U *be a unitary operator without multiplicity. Then* disc U≤2 *and* disc U=1 *iff* U *is reductive. In any case* U *has many invariant subspaces.*

PROOF. We suppose at first that U is reductive. In this case

$$\mathrm{Lat}\ U=\{\chi_e L^2(\mathbb{T},\nu): e\subset\mathbb{T}\}$$

where χ_e is the characteristic function of the set e. So the required result will be implied by the following easy lemma, whose proof we omit.

1) i.e. Θ is an analytic function in $\mathbb{D}$ whose values are contractions from $\mathbb{C}^m$ into $\mathbb{C}^n$ and whose boundary values on $\mathbb{T}$ are isometries.

LEMMA. *Let* ν *be a* σ*-finite measure on* $\mathbb{T}$, $R\subset L^1(\nu)$, dim $R<\infty$. *Then almost all functions* f *from* R *are such that*

$$e\subset\mathbb{T},\ \int_e |f|d\nu=0\Rightarrow\int_e |g|d\nu=0,\quad \forall g\in R.$$

Now we shall consider a nonreductive operator U. And we can suppose that μ_a is the Lebesgue measure. Since disc $(U_a\oplus U_s)\geq$ $\geq$disc U_a, disc U>1 will be proved if we find a two dimensional subspace of $L^2(\mathbb{T})$ cyclic for the operator of multiplication by z and containing no cyclic functions. Recall that $f\in L^2(\mathbb{T})$ is cyclic for the multiplication by z iff $f(\xi)=0$ a.e. on $\mathbb{T}$ and $\log f\notin L^1(\mathbb{T})$ (Beurling-Helson theorem, see e.g. [4]). So the subspace $R=\text{span}\ (\chi_+,\chi_-)$, where $\chi_\pm$ are the characteristic functions of semicircles $\mathbb{T}_\pm=\mathbb{T}\cap\{\xi:\pm\text{Im}\xi>0\}$ will do.

Let $R\in\text{Cyc}$ U. Then $\chi_e R\neq\{0\}$ if $\nu e\neq 0$. So using the lemma we can assert that almost all $f\in R$ satisfy $\chi_e f\neq 0$. Let $E=\text{span}\ (z^n f: n\geq 0)$ for an arbitrary f, enjoying the last property. It is clear that $E_s=L^2(\mathbb{T},\nu_s)$. So if $E_a=L^2(\mathbb{T},m)$ everything is all right. If not, $E_a=\omega H^2=\{\omega g: g\in H^2\}$ by Beurling-Helson theorem. Here $\omega\in L^\infty(\mathbb{T})$, $|\omega|=1$ m-a.e. To end the proof we need only to show that the operator $U_1=P_{E^\perp}U|E^\perp$ has many cyclic vectors in every finite dimensional cyclic subspace. But this assertion is true, because U_1 is unitarily equivalent to S^*.■

4. ISOMETRIES, COISOMETRIES, UNITARY OPERATORS AND C_o-CONTRACTIONS

In this and the following two sections we state without proof some more difficult results about the disc-calculation for some classes of operators.

4.1. PROPOSITION. disc $S_n=n+1$ *and* S_n *has many cyclic subspaces.*

4.2. UNITARY OPERATORS. Let U be a unitary operator in a separable Hilbert space H. We can think of H as of

$$H=\int_{\mathbb{T}}\oplus H(t)\,d\nu(t),$$

the spectral representation for U. The ν-a.e. defined function $r(t)\overset{\text{def}}{=}\dim H(t)$ is the local multiplicity of the spectrum of

U. The numbers

$$\mu_s=\nu_s - \text{ess sup } r(t),$$
$$\mu_a=\nu_a - \text{ess sup } r(t),$$
$$\underline{\mu}=\nu_a\text{-ess inf } r(t)$$

are the spectral multiplicities of U_s,U_a and of bilateral shift contained in U. So we can treat U as the following sum

$$U=U_s\oplus U_{ar}\oplus S_{\underline{\mu}}.$$

Here U_{ar} is the"maximal" reductive part of U_a, and $S_{\underline{\mu}}$ being the bilateral shift of multiplicity $\underline{\mu}$ (i.e. the operator of multiplication by $z^{\underline{\mu}}$ in $L^2(\mathbb{T})$) is the pure non reductive (i.e. without reductive parts) part of U. Now we can state the following

PROPOSITION.

$$\text{disc } U_s=\mu_s,$$
$$\text{disc } U_{ar}=\mu_{ar}=\mu_a-\underline{\mu},$$
$$\text{disc } S_{\underline{\mu}}=2\underline{\mu},$$
$$\text{disc } U_a=\mu_a+\underline{\mu},$$
$$\text{disc } U=\max\{\mu_s,\mu_a+\underline{\mu}\},$$

and all operators have many cyclic subspaces.

Let us make some remarks about these formulas.

Because some simple facts on polinomial approximation it is easy to show that Lat U=Lat $U_a\oplus$Lat U_s. Hence disc U= max{disc U_s, disc U_a} (see 2.11). For a reductive operator U (i.e. $\underline{\mu}=0$)we have

$$\text{disc } U=\mu_u=\max\{\mu_s,\mu_a\},$$

i.e. disc is equal to multiplicity. For a pure non reductive operator S_n we have

$$\text{disc } S_n=2\mu_{S_n}=2n,$$

i.e. disc of bilateral shift is twice more than its multiplicity. Lat S_μ is independent of Lat U_s but not of Lat U_{ar}, so we have

$$\text{disc } (U_s\oplus S_{\underline{\mu}})=\max\{\text{disc } U_s,\text{disc } S_{\underline{\mu}}\}$$

but

$$\text{disc } U_a=\text{disc}(U_{ar}\oplus S_{\underline{\mu}})=\text{disc } U_{ar}+\text{disc } S_{\underline{\mu}}.$$

4.3. C_o-CONTRACTIONS are (by definition) contractions T admitting a bounded function m, analytic in $\mathbb{D}$ with $m(T)\equiv 0$.

PROPOSITION. *If* T *is a* C_o*-contraction in a separable Hilbert space then* disc $T=\mu_T$ *and* T *has many cyclic subspaces.*

4.4. SUMS OF OPERATORS. Now we state a unifying result concerning all kinds of operators discussed above. We consider the direct sum[1)] of operators:

$$A=U\oplus S_n\oplus S_k^*\oplus T,$$

where U is unitary and T is C_o-contraction.

THEOREM.

$$\text{disc } A=\max\{\mu_s, n+\max(1,\mu_T,k+\mu_a+\underline{\mu})\}$$

and A *has many cyclic subspaces.*

All preceding propositions can be viewed as corollaries of this theorem with a special choice of A. For the proof see [3].

5. NORMAL OPERATORS

In this section we describe all normal operators in Hilbert space with disc N=1 and all normal operators satisfying disc N= $=\mu_N$ simultaneously with all their normal parts.

5.1. SOME NOTATIONS. Let λ be a finite Borel measure on $\mathbb{C}$, G be a simply connected bounded domain admitting the harmonic measure on its boundary. Then

$$\lambda_G \overset{\text{def}}{=\!=} \lambda|G+\partial_G\lambda$$

where $\partial_G\lambda$ is the part of λ absolutely continuous with respect to the harmonic measure. $H^\infty(G)$ is by definition the space of all bounded analytic functions in G with the usual sup-norm $\|\cdot\|_\infty$. The notation $\lambda_1 \ll \lambda_2$ will mean the measure λ_1 is absolutely continuous with respect to λ_2.

5.2. CRITERIA OF REDUCTIVITY for normal operators we formulate following D.Sarason [7].

Let N *be a normal operator,* E_N *be its spectral measure,* λ *be a scalar measure which is mutually absolutely continuous with* E_N. *Then the following assertions are equivalent:*

---

1) We may assume that the sum is orthogonal (see Proposition 2.5).

1) Lat N=Lat N^* (*i.e.* N *is reductive*).

2) $\nu \ll E_N$ *and* $\int z^n d\nu=0$ $\forall n\geq 0$ *imply* $\nu\equiv 0$.

3) *The set of all polynomials of* z *is weakly* $\sigma(L^\infty(\lambda), L^1(\lambda))$-*dense in* $L^\infty(\lambda)$.

There exists no bounded simply connected domain G *with the property*

$$||f||_\infty=\lambda_G\text{-ess sup}|f|, \qquad \forall f\in H^\infty(G).$$

Now we can formulate our main result on normal operators.

5.3. THEOREM. *Let* λ *be a finite measure on* $\mathbb{C}$ *with compact support. Then the following assertions are equivalent:*

1) *If* N *is normal and* $E_N\ll\lambda$, *then* disc $N=\mu_N$.

2) *If* N *is normal and* $E_N\ll\lambda$, *then* Lat N=Lat N^* (i.e. N^**is reductive*).

3) $\nu\ll\lambda$ *and* $\int z^n d\nu=0$ $\forall n\geq 0$ *imply* $\nu\equiv 0$.

4) *There exists no bounded simply connected domain* G *with the property*

$$||f||_\infty=\lambda_G\text{-ess sup}|f| \qquad \forall f\in H^\infty(G).$$

If one (and hence all) of these conditions is fulfilled, then a normal operator N *with finite spectral multiplicity has many cyclic subspaces, provided* $E_N\ll\lambda$.

5.4. COROLLARY. *Let* N *be a normal operator. Then*

1) disc N=1 *iff* $\mu_N=1$ *and* N *is reductive*.

2) N *is reductive iff* disc $N'=\mu_{N'}$ *for every normal part* $N'=N|E$, $E\in\text{Lat } N\cap\text{Lat } N^*$.

The most essential part of the theorem is the implication 1)$\Longrightarrow$2), because the implication 2)$\Longrightarrow$1) is easy and the equivalence of 2), 3) and 4) follows from 5.2.

6. COMPLETE OPERATORS, COMPACT OPERATORS, TOEPLITZ OPERATORS AND SOME OTHER CLASSES

6.1. COMPLETE OPERATOR is a linear transformation A of Banach space X with the property

$$X=\text{span}\{\text{Ker}(A-\lambda I)^n: n\geq 0,\ \lambda\in\mathbb{C}\}.$$

These operators are close (in a sense) to operators in finite dimensional spaces. Using Proposition 2.13 we analyse in [2] some characteristic of complete operator A influencing disc A. We begin with the representation (under some conditions, of course) of A as a sum of A_λ's such that for every A_λ the root manifold

$$K_\lambda \overset{\text{def}}{=\!=} \bigcup\{\text{Ker}(A-\lambda I)^n : n\geq 1\}$$

is dense in the domain of A_λ. Suppose one of these manifolds (say K_0) is dense in X and denote

$$\text{Lift } A=\{x^0 : x^0\in\text{Ker } A;\ \exists\{x^n\}_{n\geq 0} : Ax^{n+1}=x^n, n\geq 0\}.$$

Under this assumption disc A depends rather on the dimension of Lift A, than on the whole Ker A, as it could be expected (by analogy with the case dim $X<\infty$).

In [2] the following theorem is proved.

6.2. THEOREM. *If* clos $K_0=X$, *then*

$$\text{disc } A\geq\text{dim Lift } A, \tag{5}$$

$$\mu_A=\max\{1, \text{dim Ker } A^*\}.$$

If A *is semi-Fredholm and* clos $K_0=X$, *then* ind $A\geq 0$[1] *and* dim Lift A=ind A.

[2] contains a strengthened variant of (5), some cases of equality in (5), corollaries of the theorem. Some of them we formulate here.

6.3. TOEPLITZ OPERATOR $T_{\bar\phi}$ is (by definition) the following operator in H^2:

$$T_{\bar\phi}f=P_+\bar\phi f, \qquad f\in H^2,$$

where $\phi\in H^\infty(\mathbb{D})$ and P_+ is the orthogonal projection L^2 onto H^2. Then

$$\text{disc } T_{\bar\phi}\geq\sup_{\lambda\in\mathbb{C}}\ \text{card}\{\xi : \xi\in\mathbb{D}, \phi(\xi)=\lambda\}.$$

We call ϕ *n-sheeted*, if the equation $\phi(\xi)=\lambda$ has exactly n roots (counted with their multiplicities) for every $\lambda\in\phi(\mathbb{D})$. We shall write $\Omega\in(DS)$, where Ω is a simply connected domain in $\mathbb{C}$, if polynomials are weakly dense in $H^\infty(\Omega)$. Now we can state the following proposition.

Suppose ϕ *is* n*-sheeted and* $\phi(\mathbb{D})\in(DS)$. *Then* disc $T_{\bar\phi}=n$ *and* $T_{\bar\phi}$ *has many cyclic subspaces.*

[1] ind A $\overset{\text{def}}{=\!=}$ dim Ker A-dim Ker A*.

6.4. A COMPACT OPERATOR K with the property

$$X=\text{span}[\,\text{Ker}(K-\lambda I)^n : n\geq 0, \lambda\in\mathbb{C}\setminus\{0\}]$$

satisfies

$$\text{disc } K = \sup_{\lambda\neq 0} \dim \text{Ker}(K-\lambda I).$$

6.5. A STRICTLY UPPER TRIANGULAR OPERATOR A in the space $l^2(\mathbb{C}^n)$ has disc $A\geq n$, if the matrix elements $A_{k,k+1}$ are invertible. This occurs, for instance, for a backward weighted shift if its weights are invertible.

7. UNSOLVED PROBLEM

7.1. IS IT TRUE, THAT EVERY OPERATOR HAS MANY CYCLIC SUBSPACES? For every operator, for which we can compute its disc, the answer is "yes".

7.2. IS THE disc INVARIANT UNDER QUASIMILARITY? The spectral multiplicity is invariant under quasisimilarity. We conjecture it is not true for the disc, but we do not know any example.

7.3. IS THE INEQUALITY

$$\text{disc } (A\oplus B)\leq \text{disc } A+\text{disc } B$$

TRUE FOR ARBITRARY OPERATORS A AND B? (See 2.12). If there exists a non reductive operator N such that:

1) $\mu_N=1$,

2) disc $N>1$,

3) $G\setminus\sigma(N)$ is connected (where G is domain from the 4th assertion of 5.2 and $\sigma(N)$ is the spectrum of N), then the answer is "no".

7.4. LET A BE SEMI-FREDHOLM AND ind $A<0$. IS THEN disc $A\geq |\text{ind } A|+1$? This question is naturally connected with the inequality disc $S_n\geq n+1$ and with a theorem of J.Wermer [9]. By this theorem every operator A with Ker $A=\{0\}$ and $AX=\text{clos } AX$ is similar to the operator of multiplication by z in a space of analytic functions.

7.5. IS disc $A\geq 2$ FOR EVERY SUBNORMAL NONNORMAL OPERATOR A? Some more questions about a normal operator N and a subnormal operator A without normal parts:

is $\mu_{A^*}=1$?

is disc $A^*=\mu_A$?

How disc N can be computed if we know the spectral measure $\mathcal{E}_N$?

When is disc $N=\mu_N$?

In connection with the last two questions see §5.

7.6. IS disc $T_{\bar\phi}=\max_\lambda \text{card}\{\xi:\phi(\xi)=\lambda\}$, FOR A FUNCTION ϕ, $\phi\in H^\infty$ AND SMOOTH IN clos $\mathbb{D}$? It would be interesting, of course, to compute the spectral multiplicity of $T_{\bar\phi}$, both for $\phi\in H^\infty$ and $\phi\in \overline{H^\infty}$ and for more general ϕ (cf.[1,2]).

REFERENCES

1. Васюнин, В.И.; Никольский, Н.К.: Управляющие подпространства минимальной размерности. Элементарное введение. Discotheca, Записки научн. семинаров ЛОМИ, Ленинград, 113(1981), 41-75.

2. Васюнин, В.И.; Никольский, Н.К.: Управляющие подпространства минимальной размерности. Влияние корневых векторов. Discotheca, Препринт ЛОМИ, P-4-81, Ленинград, 1981, 1-45.

3. Васюнин, В.И.; Никольский, Н.К.: Управляющие подпространства минимальной размерности. Унитарные и модельные операторы. Discotheca, Препринт ЛОМИ, P-5-81, Ленинград, 1981.

4. Никольский, Н.К.: Лекции об операторе сдвига, Москва, "Наука", 1980.

5. Feintuch, A: On single input controllability for infinite dimensional linear systems, *J.Math.Anal.Appl.* 62(1978), 538-546.

6. Heymann, M.: On the input and output reducibility of multivariable linear systems, *IEEE Trans.Aut.Control* AC-15(1970), 563-569.

7. Sarason, D.: Weak-star density of polynomials, *J.Reine Angew. Math.* 252(1972), 1-15.

8. Wermer, J.: On invariant subspaces of normal operators, *Proc.Amer. Math.Soc.* 3(1952), 270-277.

9. Wermer, J.: On restrictions of operators, *Proc.Amer.Math.Soc.* 4(1953), 860-865.

N.K.Nikolskii and V.I.Vasjunin
191011 Leningrad
Fontanka 27, LOMI
USSR.

DERIVATIONS OF C^*-ALGEBRAS WHICH ARE INVARIANT UNDER AN AUTOMORPHISM GROUP. II

C.Peligrad

1. INTRODUCTION

By a C^*-dynamical system we mean a triple (B,G,β) consisting of a C^*-algebra B, a locally compact group G and a continuous homomorphism β of G into the group Aut(B) of *-automorphism of B equipped with the topology of pointwise convergence.

In [5] we studied derivations of B commuting with β. Namely we showed there that if B has a unit and G is compact and ergodically acting then every derivation of B commuting with β is a generator. We also considered there derivations of crossed products of simple C^*-algebras with abelian discrete groups of automorphisms.

In this paper we obtain further results in this direction. Theorem 2.5 below shows that if G is compact, and $\delta:\mathcal{D}(\delta)\to B$, $(\mathcal{D}(\delta)\subset B)$ is a derivation of B which commutes with β such that $\sum_{\hat{\pi}\in\hat{G}} B(\hat{\pi})\subset\mathcal{D}(\delta)$ then δ is a generator. Theorem 3.7 is a significant improvment of [5, Theorem 3.5].

We also consider derivations satisfying a condition which is weaker than the commutation condition (see the condition (*) below).

Finally we give the corresponding results for the more general case of Lie algebras of derivations commuting with compact actions.

2. DERIVATIONS COMMUTING WITH COMPACT AUTOMORPHISMS GROUPS

Let (B,G,β) be a C^*-dynamical system. Assume that G is compact.

For $f \in L^1(G)$ we denote by $\beta(f)$ the operator on B defined by

$$\beta(f)b = \int_G f(g)\beta_g(f)dg .$$

Let π be an irreducible unitary representation of G, $\hat{\pi}\in\hat{G}$ its unitary equivalence class, and $\chi_{\hat{\pi}}$ its normalised character $\chi_{\hat{\pi}}(g) = (\dim \pi)\mathrm{Tr}(\pi g^{-1})$ where Tr is the usual trace on the Hilbert space of dimension $\dim\pi$. Then $\beta(\chi_{\hat{\pi}})$ is a bounded projection of B onto a norm closed subspace $B(\hat{\pi})$ of B, called the spectral subspace of $\hat{\pi}$ in B.

We denote $\beta(\chi_{\hat{\pi}})$ by $\beta(\hat{\pi})$. The trivial one dimensional representation will be denoted by π_o.

It is well known (and easy to show) that there exists a one to one correspondence between the states φ of $B(\hat{\pi}_o)$ and the G--invariant states $\tilde{\varphi}$ of B given by $\tilde{\varphi}(b) = \varphi(\beta(\hat{\pi}_o)(b))$, $b \in B$.

Let φ be a state of $B(\hat{\pi}_o)$ and $\tilde{\varphi}$ be the corresponding G-invariant state of B. Consider the GNS representation of B, $(H_{\tilde{\varphi}}, \pi_{\tilde{\varphi}}, \xi_{\tilde{\varphi}})$ determined by $\tilde{\varphi}$. It is obvious that $\ker\pi_{\tilde{\varphi}}$ is a G-invariant ideal of B for all $\tilde{\varphi}$. For each $\hat{\pi}\in\hat{G}$ denote by $H_{\tilde{\varphi}}^{\hat{\pi}}$ the closure (in $H_{\tilde{\varphi}}$) of $\pi_{\tilde{\varphi}}(B(\hat{\pi}))\xi_{\tilde{\varphi}}$. Denote, also by $M_{\tilde{\varphi}}$ the σ-weak closure of $\pi_{\tilde{\varphi}}(B)$ in $B(H_{\tilde{\varphi}})$.

The following lemma is well known.

LEMMA 2.1. *There exists a representation* $\beta^{\tilde{\varphi}}$ *of* G *as a group of *-automorphism of* $M_{\tilde{\varphi}}$ *such that* $(M_{\tilde{\varphi}}, G, \beta^{\tilde{\varphi}})$ *is a* W**-dynamical system and* $\beta_g^{\tilde{\varphi}}(\pi_{\tilde{\varphi}}(a)) = \pi_{\tilde{\varphi}}(\beta_g(a))$ *for all* $a\in B$. *Moreover, we have:* $M_{\tilde{\varphi}}(\hat{\pi}) = \pi_{\tilde{\varphi}}(B(\hat{\pi}))$ *for all* $\hat{\pi}\in\hat{G}$.

PROOF. Let $U_g^{\tilde{\varphi}}(\pi_{\tilde{\varphi}}(a)\xi_{\tilde{\varphi}}) = \pi_{\tilde{\varphi}}(\beta_g(a))\xi_{\tilde{\varphi}}$, $g\in G$, $a\in B$. Then, since $\tilde{\varphi}$ is G invariant, U_g has a unitary extension on $H_{\tilde{\varphi}}$ and $g \to U_g^{\tilde{\varphi}}$ is strongly continuous. Next we put $\beta_g^{\tilde{\varphi}}(\pi_{\tilde{\varphi}}(a)) = U_g^{\tilde{\varphi}}\pi_{\tilde{\varphi}}(a)U_g^{\tilde{\varphi}-1}$, $g\in G$, $a\in B$. Then the first part of the lemma is proved. The second part follows from the fact that the projections $\beta^{\tilde{\varphi}}(\hat{\pi})$ of $M_{\tilde{\varphi}}$ onto $M_{\tilde{\varphi}}(\hat{\pi})$ are σ-weakly continuous.

Now, we consider a symmetric, closed, densely defined derivation $\delta:\mathcal{D}(\delta)\to B$, $(\mathcal{D}(\delta)\subset B)$.

REMARK 2.2. If $\beta_g\delta\subset\delta\beta_g$ for all $g\in G$, then it is easy to show that $\beta(\hat{\pi})\delta\subset\delta\beta(\hat{\pi})$ for all $\hat{\pi}\in\hat{G}$. The following lemma is an easy adaptation of [2, Proposition 3.2.28].

LEMMA 2.3. *Let* $\delta:\mathcal{D}(\delta)\to B$, $(\mathcal{D}(\delta)\subset B)$ *be a symmetric, densely defined derivation, and let* φ *be a state of* B. *Let also* $(H_\varphi,\pi_\varphi,\xi_\varphi)$ *be the* GNS *representation of* B *determined by* φ. *Then the following conditions are equivalent:*

1) $$|\varphi(\delta(a))|\le L(\varphi(a^*a)+\varphi(aa^*))^{1/2}$$

for all $a\in\mathcal{D}(\delta)$ *and some* $L\ge 0$.

2) *There exists a symmetric operator* h_φ *on* H_φ *such that*

$$\mathcal{D}(h_\varphi)=\pi_\varphi(\mathcal{D}(\delta))\xi_\varphi$$

and

$$\pi_\varphi(\delta(a))\xi=i[h_\varphi,\pi_\varphi(a)]\xi$$

for all $a\in\mathcal{D}(\delta)$ *and* $\xi\in\mathcal{D}(h_\varphi)$.

Now, let (B,G,β) be as above and let $\delta:\mathcal{D}(\delta)\to B$, $(\mathcal{D}(\delta)\subset B)$, be a symmetric, closed, densely defined derivation such that δ commutes with β and $B(\hat{\pi}_o)\subset\mathcal{D}(\delta)$. Let φ be a state of $B(\hat{\pi}_o)$ and let $\tilde{\varphi}=\varphi\circ\beta(\hat{\pi}_o)$ be the corresponding G-invariant state of B.

Since $B(\hat{\pi}_o)\subset\mathcal{D}(\delta)$, by a well known result of Sakai, it follows that $\delta|B(\hat{\pi}_o)$ is bounded. Therefore, as it is also known, $\delta|B(\hat{\pi}_o)$ is σ-weakly continuous. Hence, there exists a self-adjoint $h\in B(\hat{\pi}_o)''$ such that $\delta|B(\hat{\pi}_o)=ad(h)|B(\hat{\pi}_o)$.

A similar but simpler argument with that of the preceding lemma shows that:

$$|\varphi(\delta(a))|\le L(\varphi(a^*a)+\varphi(aa^*))^{1/2}$$

for all $a\in B(\hat{\pi}_o)$ and some $L\ge 0$.

Using the fact that $\beta(\hat{\pi}_o)\delta\subset\delta\beta(\hat{\pi}_o)$ and the properties of the conditional expectation $\beta(\hat{\pi}_o)$ (see for exemple [7, Theorem 3.4]) we have

$$|\tilde{\varphi}(\delta(a))|\le L(\tilde{\varphi}(a^*a)+\tilde{\varphi}(aa^*))^{1/2}$$

for all $a\in\mathcal{D}(\delta)$. Therefore, by the preceding lemma, we may consider the following derivation on $\pi_{\tilde{\varphi}}(B)$

$$\delta^{\tilde{\varphi}}(\pi_{\tilde{\varphi}}(a))=\pi_{\tilde{\varphi}}(\delta(a))=i[h_{\tilde{\varphi}},\pi_{\tilde{\varphi}}(a)] \qquad a\in\mathcal{D}(\delta).$$

LEMMA 2.4. *Let* (B,G,β) *be a* C^**-dynamical system with* G *compact. Let also* $\delta:\mathcal{D}(\delta)\to B$, $(\mathcal{D}(\delta)\subset B)$ *be a symmetric, closed, densely defined derivation which commutes with* β. *Assume that* $\sum_{\hat{\pi}\in\hat{G}} B(\hat{\pi})\subset\mathcal{D}(\delta)$. *If* φ *is a state of* $B(\hat{\pi}_o)$, *then there exist an essentially selfadjoint operator* $h_{\tilde{\varphi}}$ *on* $H_{\tilde{\varphi}}$, *a* $*$*-subalgebra* $\mathcal{D}\subset M_{\tilde{\varphi}}$ *and a derivation* δ_1 *on* $\mathcal{D}$ *which extends* $\delta^{\tilde{\varphi}}$ *such that:*

(i) $\pi_{\tilde{\varphi}}(\mathcal{D}(\delta))\subset\mathcal{D}$,

(ii) $\mathcal{D}(h_{\tilde{\varphi}})=\mathcal{D}\xi_{\tilde{\varphi}}$,

(iii) $\delta_1(a_{\tilde{\varphi}})\xi=i[h_{\tilde{\varphi}},a_{\tilde{\varphi}}]\xi$, $a_{\tilde{\varphi}}\in\mathcal{D}$, $\xi\in\mathcal{D}(h_{\tilde{\varphi}})$.

Moreover we have $\|\pi_{\tilde{\varphi}}[(1+\alpha\delta)(a)]\|\geq\|\pi_{\tilde{\varphi}}(a)\|$ *for* $\alpha\in\mathbb{R}$, $a\in\mathcal{D}(\delta)$.

PROOF. By the remark preceding this lemma, we have $\delta^{\tilde{\varphi}}(\pi_{\tilde{\varphi}}(a))\xi=i[h'_{\tilde{\varphi}},\pi_{\tilde{\varphi}}(a)]\xi$ for all $a\in\mathcal{D}(\delta)$ and some symmetric operator $h'_{\tilde{\varphi}}$ on $H_{\tilde{\varphi}}$, $\xi\in\mathcal{D}(h'_{\tilde{\varphi}})$. Then, it follows that $\delta^{\tilde{\varphi}}$ is σ-weakly closable on $M_{\tilde{\varphi}}$.

We denote its σ-weak closure by $\overline{\delta}^{\tilde{\varphi}}$. Since $B(\hat{\pi}_o)\subset\mathcal{D}(\delta)$ it follows that $M_{\tilde{\varphi}}(\hat{\pi}_o)\subset\mathcal{D}(\overline{\delta}^{\tilde{\varphi}})$ and that there exists a $h\in M_{\tilde{\varphi}}(\hat{\pi}_o)=\overline{\pi_{\tilde{\varphi}}(B(\hat{\pi}_o))}^{\sigma}$ (by Lemma 2.1) such that $\overline{\delta}^{\tilde{\varphi}}(a_{\tilde{\varphi}})=i[h,a_{\tilde{\varphi}}]$ for all $a_{\tilde{\varphi}}\in M_{\tilde{\varphi}}(\hat{\pi}_o)$. In particular $\overline{\delta}^{\tilde{\varphi}}(h)=0$.

Since obviously $\mathcal{D}(\overline{\delta}^{\tilde{\varphi}})$ is a $*$-subalgebra of $M_{\tilde{\varphi}}$, we may consider the $*$-subalgebra $\mathcal{D}\subset\mathcal{D}(\overline{\delta}^{\tilde{\varphi}})$ generated by h and $\pi_{\tilde{\varphi}}(\mathcal{D}(\delta))$ and put $\delta_1=\overline{\delta}^{\tilde{\varphi}}|\mathcal{D}$. Then $\pi_{\tilde{\varphi}}(\mathcal{D}(\delta))\subset\mathcal{D}$ and δ_1 extends $\delta^{\tilde{\varphi}}$.

Let $\delta_o^{\tilde{\varphi}}=\delta_1-\operatorname{ad}(ih)$. Then $\delta_o^{\tilde{\varphi}}$ is a symmetric, σ-weakly closable, densely defined derivation on $\mathcal{D}$ such that

$$\delta_o^{\tilde{\varphi}}(h)=0 \qquad \text{and} \qquad \delta_o^{\tilde{\varphi}}|\pi_{\tilde{\varphi}}(B(\hat{\pi}_o))=0.$$

Then, as in the proof of [2, Proposition 3.2.58] it can be shown that $\delta_o^{\tilde{\varphi}}$ is implemented by the following symmetric operator:

$$h_o a_{\tilde\varphi}\xi_{\tilde\varphi}=\delta_o^{\tilde\varphi}(a_{\tilde\varphi})\xi_{\tilde\varphi} \quad , \quad a_{\tilde\varphi}\in\mathcal{D} .$$

(since $\tilde\varphi(\delta_o^{\tilde\varphi}(a_{\tilde\varphi}))=0$ for all $a_{\tilde\varphi}\in\mathcal{D}$).

Next, we prove that $\sum_{\hat\pi\in\hat G}\pi_{\tilde\varphi}(B(\hat\pi))\xi_{\tilde\varphi}$ is a set of analytic elements for h_o. Let $a\in B(\hat\pi)$, $\hat\pi\in\hat G$. Since $\delta_1(h)=\delta_o^{\tilde\varphi}(h)=0$, it can be easily verified by recurrence the following relation:

$$(\delta_o^{\tilde\varphi})^n(\pi_{\tilde\varphi}(a))=\sum_{k=0}^{n}(-1)^k C_n^k[\mathrm{ad}(ih)]^k((\delta_1)^{n-k}(\pi_{\tilde\varphi}(a))) .$$

Since $B(\hat\pi)\subset\mathcal{D}(\delta)$ and $\pi_{\tilde\varphi}(\mathcal{D}(\delta))\subset\mathcal{D}$, it follows that $\delta_1|\pi_{\tilde\varphi}(B(\hat\pi))$ is bounded. Let $d\geq 0$ be the norm of this restriction. Then we have:

$$||(\delta_o^{\tilde\varphi})^n(\pi_{\tilde\varphi}(a))||\leq(d+2||h||)^n||\pi_{\tilde\varphi}(a)|| \quad , \quad n\in\mathbb{N} .$$

Therefore

$$||h_o^n\pi_{\tilde\varphi}(a)\xi_{\tilde\varphi}||\leq(d+2||h||)^n||\pi_{\tilde\varphi}(a)|| .$$

It follows that $\pi_{\tilde\varphi}(a)\xi_{\tilde\varphi}$ is analytic for h_o. Hence $\sum_{\hat\pi\in\hat G}\pi_{\tilde\varphi}(B(\hat\pi))\xi_{\tilde\varphi}$ is a dense set of analytic elements for h_o , and so by Nelson's theorem h_o is essentially selfadjoint. Therefore $h_{\tilde\varphi}=h_o+h$ is essentially selfadjoint. The operator $h_{\tilde\varphi}$ and the derivation δ_1 satisfy obviously the conditions (i)-(iii) of the lemma. The last conclusion of the lemma follows from [2, Corollary 3.2.56].

THEOREM 2.5. *Let* (B,G,β) *be a* C^**-dynamical system with* G *compact and let* δ *be a norm densely defined, norm-closed, symmetric derivation of* B. *Assume that*

1) $\beta_g\delta\subset\delta\beta_g$, $\quad g\in G$,

2) $\sum_{\hat\pi\in\hat G}B(\hat\pi)\subset\mathcal{D}(\delta)$.

Then δ *is a generator.*

PROOF. Obviously $\sum_{\hat\pi\in\hat G}B(\hat\pi)$ is a dense set of analytic elements for δ. On the other hand by Lemma 2.4 we have

$$||(1+\alpha\delta)(a)||=\sup_{\varphi}||\pi_{\tilde{\varphi}}((1+\alpha\delta)(a))||\geq\sup_{\varphi}||\pi_{\tilde{\varphi}}(a)||=||a||,$$

$a\in\mathcal{D}(\delta)$ where the supremum is taken over all states φ of $B(\hat{\pi}_o)$. Therefore δ satisfies the conditions A1), B2), C1) of [2, Theorem 3.2.50] and so δ is a generator.

Now we consider the following condition on δ which is implied by the commutation with β:

$$(*)\qquad \begin{aligned}&\beta_g\mathcal{D}(\delta)\subset\mathcal{D}(\delta),\ g\in G \text{ and there exists } M>0 \text{ such that}\\ &||\beta_g\delta(a)-\delta\beta_g(a)||\leq M||a||,\ g\in G,\quad a\in\mathcal{D}(\delta).\end{aligned}$$

PROPOSITION 2.6. *Let* (B,G,β) *be as above. Suppose in addition that* B *is separable. Let* $\delta:\mathcal{D}(\delta)\to B$ *be a norm-densely defined, norm closed symmetric derivation which satisfies the condition* (*). *Then there exists a derivation* δ_1 *which commutes with* β *and a bounded derivation* δ_2 *such that* $\delta=\delta_1+\delta_2$.

PROOF. Let $a\in\mathcal{D}(\delta)$. Then by condition (*) the orbit $\mathcal{O}(a)=$ $=\{\beta_g(a)\}_{g\in G}$ is a compact subset of $\mathcal{D}(\delta)$. Since δ is norm closed, it follows that the set $\{b\oplus\delta(b)\,|\,b\in\mathcal{O}(a)\}$ is norm closed in $B\oplus B$. Since B is separable, we can apply the measurable choice theorem of von Neumann (see for example [9]). It follows that the function $g\to\delta(\beta_g(a))$ is m-measurable in the sense of [1] (where m is the Haar measure of G).

By [1] the function $g\to\beta_{g^{-1}}\delta(\beta_g(a))$ is m-measurable. From condition (*) it follows easily that this function is bounded (in norm) and thus it is integrable.

Put $\delta_1(a)=\int_G\beta_{g^{-1}}\delta(\beta_g(a))dg$ and $\delta_2(a)=\delta(a)-\delta_1(a)$ which completes the proof.

COROLLARY 2.7. *Let* (B,G,β) *be a* C*-*dynamical system with* G-*-compact and* B *separable. Let* δ *be a norm closed norm densely defined, symmetric derivation which satisfies* (*). *If* $\sum_{\hat{\pi}\in\hat{G}}B(\hat{\pi})\subset\mathcal{D}(\delta)$ *then* δ *is a generator.*

PROOF. The proof follows from Proposition 2.5, Proposition

2.6, and Phillips's bounded perturbation theorem.

3. DERIVATIONS OF CROSSED PRODUCTS

In this section we consider the case of C^*-crossed products. Let (A,Γ,α) be a C^*-dynamical system with Γ discrete. Suppose $A\subset L(H)$ for some Hilbert space H. For notations and results from the theory of crossed products we refer to [4]. However for the sake of completeness we give here some notations and definitions. We shall assume that A is unital.

Let $C_{\infty}(\Gamma)$ be the space of all A-valued functions on Γ with compact support.

The reduced crossed product of A and Γ, denoted $C_r^*(A,\Gamma,\alpha)$ is defined as the C^*-algebra generated by the operators $\tilde{\rho}(\varphi)$ on $L^2(\Gamma,H)$:

$$\tilde{\rho}(\varphi)f(\gamma)=\sum_{p\in\Gamma}\alpha_{\gamma^{-1}}(\varphi(p))f(p^{-1}\gamma) \tag{1}$$

for $\varphi\in C_{\infty}(\Gamma)$, $f\in L^2(\Gamma,H)$ and $\gamma\in\Gamma$.

In particular, if $p\in\Gamma$ and $\varphi_p(\gamma)=\begin{cases}1 & \text{if } \gamma=p\\ 0 & \text{if } \gamma\neq p\end{cases}$ we put $\tilde{\rho}(\varphi_p)=\overline{\lambda}_p$.

Also if $a\in A$ and $\varphi(\gamma)=\begin{cases}a & \text{if } \gamma=e\\ 0 & \text{if } \gamma\neq e\end{cases}$ we put

$$\tilde{\rho}(\varphi)=\tilde{\rho}(a)\ . \tag{2}$$

Then it is easy to see that $C_r^*(A,\Gamma,\alpha)$ is generated by the direct sum $\sum_{\gamma\in\Gamma}\tilde{\rho}(A)\overline{\lambda}_{\gamma}$.

Now, we define the dual action β on $C_r^*(A,\Gamma,\alpha)$. Let W be the following unitary operator on $L^2(\Gamma\times\Gamma,H)$

$$Wf(\gamma,p)=f(\gamma,\gamma p)\ . \tag{3}$$

Let also $C_r^*(\Gamma)$ denotes the C^*-algebra generated by the left regular representation of $C_{\infty}(\Gamma)$ on $L^2(\Gamma)$. Define the map β from $C_r^*(A,\Gamma,\alpha)$ into $L(L^2(\Gamma\times\Gamma,H)$ by:

$$\beta(a)=W^*(a\otimes 1)W \qquad a\in C_r^*(A,\Gamma,\alpha)$$

and a map β_{Γ} from $C^*(\Gamma)$ into $L(L^2(\Gamma\times\Gamma,H)$ by

$$\beta_\Gamma(a)=W_\Gamma^*(a\otimes 1)W_\Gamma \qquad a\in C_r^*(\Gamma)$$

where $W_\Gamma\xi(\gamma,p)=\xi(\gamma,\gamma p)$ if $\xi\in L^2(\Gamma\times\Gamma)$. Then we have:

$$(4)\qquad (\beta\otimes\iota)(\beta(a))=(\iota\otimes\beta_\Gamma)(\beta(a)) \qquad a\in C_r^*(A,\Gamma,\alpha)\ ,$$

$$(5)\qquad \beta(\tilde\rho(a))=\tilde\rho(a)\otimes I \qquad a\in A,$$

$$(6)\qquad \beta(\bar\lambda_\gamma)=\bar\lambda_\gamma\otimes\lambda_\gamma \qquad \gamma\in\Gamma$$

(here $\lambda_\gamma\in L(L^2(\Gamma))$ is given by $(\lambda_\gamma\xi)(p)=\xi(\gamma^{-1}p)$. For each $\gamma\in\Gamma$ define the functional θ_γ on $C_{00}(\Gamma)$ by $\theta_\gamma(f)=f(\gamma)$. It is easy to see that θ_γ has a bounded extension to all of $C_r^*(\Gamma)$ which will be denoted with θ_γ too.

Also, for each $\gamma\in\Gamma$ define the following map

$$P_\gamma: C_r^*(A,\Gamma,\alpha)\underset{alg}{\otimes} C_r^*(\Gamma)\rightarrow C^*(A,\Gamma,\alpha)$$

by

$$P_\gamma(\sum_i a_i\otimes f_i)=\sum_i\theta_\gamma(f_i)a_i\ .$$

By [8], P_γ has a bounded extension to $C_r^*(A,\Gamma,\alpha)\otimes C_r^*(\Gamma)$. Further, we denote $C_r^*(A,\Gamma,\alpha)=B$.

Let $\gamma\in\Gamma$. Define the following map:

$$q_\gamma: B\underset{alg}{\otimes} C_r^*(\Gamma)\underset{alg}{\otimes} C_r^*(\Gamma)\rightarrow B\underset{alg}{\otimes} C_r^*(\Gamma)$$

$$q_\gamma(\sum b_i\otimes f_i\otimes g_i)=\sum\theta_\gamma(g_i)b_i\otimes f_i\ .$$

Then q_γ has a bounded extension to $B\underset{min}{\otimes} C_r^*(\Gamma)\underset{min}{\otimes} C_r^*(\Gamma)$. For $\gamma=e$, these mappings were considered in [4].

The following lemma is similar with [4, Lemma 2.7] and we omit its proof.

LEMMA 3.1. *For* $x\in B\underset{min}{\otimes} C_r^*(\Gamma)$ *we have*

a) $\beta\circ P_\gamma(x)=q_\gamma(\beta\otimes\iota)(x)$

b) $q_\gamma(\iota\otimes\beta_\Gamma)(x)=P_\gamma(x)\otimes\lambda_\gamma$.

Exactly as in [4, Lemma 2.8], from the preceding lemma it follows:

LEMMA 3.2. $(P_\gamma \circ \beta)(B) = \tilde{\rho}(A)\overline{\lambda}_\gamma$.

Let $\delta : \mathcal{D}(\delta) \to B$, $(\mathcal{D}(\delta) \subset B)$ be a norm closed, norm densely defined, symmetric derivation. Then $\delta \otimes \iota$ is a densely defined symmetric derivation on $B \underset{min}{\otimes} C_r^*(\Gamma)$. Since the algebraic tensor product $B' \otimes C_r^*(\Gamma)'$ of the conjugate spaces is total in $[B \underset{min}{\otimes} C_r^*(\Gamma)]'$ (see for example [6 Corollary 4.20]) it follows that $\delta \otimes \iota$ is closable. Denote by $\delta \overline{\otimes} \iota$ its minimal closed extension. From now on we assume that A is unital.

LEMMA 3.3. *Let $\delta : \mathcal{D}(\delta) \to B$, $(\mathcal{D}(\delta) \subset B)$ be a norm closed, norm densely defined, symmetric derivation. Assume that:*

1) $\tilde{\rho}(A) \subset \mathcal{D}(\delta)$
2) $\beta\delta \subset (\delta \overline{\otimes} \iota)\beta$.

Then $\tilde{\rho}(A)\overline{\lambda}_\gamma \subset \mathcal{D}(\delta)$ *for every* $\gamma \in \Gamma$, *and* $\delta(\tilde{\rho}(A)\overline{\lambda}_\gamma) \subset \tilde{\rho}(A)\overline{\lambda}_\gamma$.

PROOF. We prove first that $\tilde{\rho}(A)\overline{\lambda}_\gamma \cap \mathcal{D}(\delta)$ is dense in $\tilde{\rho}(A)\overline{\lambda}_\gamma$, $(\gamma \in \Gamma)$. Obviously $P_\gamma(\mathcal{D}(\delta) \underset{alg}{\otimes} C_r^*(\Gamma)) \subset \mathcal{D}(\delta)$. Further, for $\Sigma a_i \otimes f_i \in \mathcal{D}(\delta) \underset{alg}{\otimes} C_r^*(\Gamma)$ we have

$$P_\gamma(\delta \otimes \iota(\Sigma a_i \otimes f_i)) = \Sigma \theta_\gamma(f_i)\delta(a_i) = \delta P_\gamma(\Sigma a_i \otimes f_i) .$$

Then, it follows that $P_\gamma \circ (\delta \overline{\otimes} \iota) = \delta \circ P_\gamma$ on $\mathcal{D}(\delta \overline{\otimes} \iota)$.

On the other hand since $\beta\delta \subset (\delta \overline{\otimes} \iota)\beta$, we have for $a \in \mathcal{D}(\delta)$:

$$\text{(6)} \qquad \delta \circ P_\gamma(\beta(a)) = P_\gamma((\delta \overline{\otimes} \iota)(\beta(a))) = P_\gamma(\beta(\delta(a))) \in \tilde{\rho}(A)\overline{\lambda}_\gamma .$$

Since P_γ and β are continuous it follows that $P_\gamma(\beta(\mathcal{D}(\delta)))$ is dense in $\tilde{\rho}(A)\overline{\lambda}_\gamma$. Now let $A_\gamma = \{a \in \tilde{\rho}(A) \mid a\overline{\lambda}_\gamma \in \mathcal{D}(\delta)\}$. Since $\tilde{\rho}(A) \subset \mathcal{D}(\delta)$ it follows that A_γ is a dense ideal in $\tilde{\rho}(A)$. Therefore $A_\gamma = \tilde{\rho}(A)$, so $\tilde{\rho}(A)\overline{\lambda}_\gamma \subset \mathcal{D}(\delta)$ and by (6) $\delta(\tilde{\rho}(A)\overline{\lambda}_\gamma) \subset \tilde{\rho}(A)\overline{\lambda}_\gamma$.

Let φ be a state of A. Then $\tilde{\varphi} = \varphi \circ P_e$ ($e \in \Gamma$ is the unit) is a state of B. Let $(H_{\tilde{\varphi}}, \pi_{\tilde{\varphi}}, \xi_{\tilde{\varphi}})$ be the GNS representation of B determined by $\tilde{\varphi}$.

We denote by $H^{\gamma}_{\tilde{\varphi}}$ $(\gamma\varepsilon\Gamma)$ the closure of $\pi_{\tilde{\varphi}}(\tilde{\rho}(A)\overline{\lambda}_{\gamma})\xi_{\tilde{\varphi}}$ in $H_{\tilde{\varphi}}$. Then $\pi_{\tilde{\varphi}}(\tilde{\rho}(A))$ may be viewed as a C^*-algebra of operators on $H^{e}_{\tilde{\varphi}}$. Let $\alpha^{\tilde{\varphi}}_{\gamma}(\pi_{\tilde{\varphi}}(\tilde{\rho}(a)))=\pi_{\tilde{\varphi}}(\overline{\lambda}_{\gamma})\pi_{\tilde{\varphi}}(a)\pi_{\tilde{\varphi}}(\overline{\lambda}_{\gamma^{-1}})$. Therefore $(\pi_{\tilde{\varphi}}(\rho(A)),\Gamma,\alpha^{\tilde{\varphi}})$ is a C^*-dynamical system (on $H^{e}_{\tilde{\varphi}}$).

LEMMA 3.4. (i) $H_{\tilde{\varphi}}\simeq L^2(\Gamma,H^{e}_{\tilde{\varphi}})$;

(ii) $\pi_{\tilde{\varphi}}(B)\simeq C^*_r(\pi_{\tilde{\varphi}}(\tilde{\rho}(A)),\Gamma,\alpha^{\tilde{\varphi}})$;

(iii) $\overline{\pi_{\tilde{\varphi}}(B)}^{\sigma}\simeq W^*_r(\overline{\pi_{\tilde{\varphi}}(\tilde{\rho}(A))}^{\sigma},\Gamma,\alpha^{\tilde{\varphi}})$,

where $W^*_r(\overline{\pi_{\tilde{\varphi}}(\tilde{\rho}(A))}^{\sigma},\Gamma,\alpha^{\tilde{\varphi}})$ *is the reduced* W^**-crossed product of* $\overline{\pi_{\tilde{\varphi}}(\tilde{\rho}(A))}^{\sigma}$ *and* Γ [4].

PROOF. (i) It is easy to see that $H_{\tilde{\varphi}}\simeq\sum_{\gamma\varepsilon\Gamma}\oplus H^{\gamma}_{\tilde{\varphi}}$. Now let $\gamma\varepsilon\Gamma$. Put $U_{\gamma}(\pi_{\tilde{\varphi}}(\tilde{\rho}(a)\overline{\lambda}_{\gamma})\xi_{\tilde{\varphi}})=\pi_{\tilde{\varphi}}(\alpha_{\gamma^{-1}}(a))\xi_{\tilde{\varphi}}$. Then U_{γ} extends to a unitary operator $U_{\gamma}:H^{\gamma}_{\tilde{\varphi}}\to H^{e}_{\tilde{\varphi}}$. (ii) and (iii) follows from [4, Theorems 3 and 1] if we define

$$\beta_{\tilde{\varphi}}(a)=W^*_{\tilde{\varphi}}(a\otimes 1)W_{\tilde{\varphi}} \qquad a\varepsilon\overline{\pi_{\tilde{\varphi}}(B)}^{\sigma}$$

where $W_{\tilde{\varphi}}f(\gamma,p)=f(\gamma,\gamma p)$, $f\varepsilon L^2(\Gamma\times\Gamma,H^{e}_{\tilde{\varphi}})$.

Further, we consider the mappings $P^{\tilde{\varphi}}_{\gamma}:\overline{\pi_{\tilde{\varphi}}(B)}^{\sigma}\,\overline{\otimes}\,L(G)\to\overline{\pi_{\tilde{\varphi}}(B)}^{\sigma}$ which are analogous with the corresponding mappings P_{γ} for the case of C^*-crossed products (see the remarks before Lemma 3.1). These mappings are σ-weakly continuous (see [4, Lemma 1.1]).

LEMMA 3.5. $\overline{P}^{\tilde{\varphi}}_{\gamma}(\beta_{\tilde{\varphi}}(\overline{\pi_{\tilde{\varphi}}(B)}^{\sigma}))=\overline{\pi_{\tilde{\varphi}}(\tilde{\rho}(A)\overline{\lambda}_{\gamma})}^{\sigma}$.

PROOF. The lemma follows from Lemmas 3.4 and 3.5, and the σ-weak continuity of $\overline{P}^{\tilde{\varphi}}_{\gamma}$ and $\beta_{\tilde{\varphi}}$.

By Lemma 2.3 we may consider the derivation $\delta^{\tilde{\varphi}}$ on $\pi_{\tilde{\varphi}}(B)$.

LEMMA 3.6. *Keep the above notations. Let* $\delta:\mathcal{D}(\delta)\to B$, $(\mathcal{D}(\delta)\subset B)$ *be a symmetric, closed, densely defined derivation which commutes with* β (i.e. $\beta\delta\subset(\delta\bar{\otimes}\iota)\beta$). *Assumed that* $\tilde{\rho}(A)\subset\mathcal{D}(\delta)$. *Let* φ *be a state of* $\tilde{\rho}(A)$ *and let* $\tilde{\varphi}=\varphi\circ P_e$ *be the corresponding state of* B. *Consider the* GNS *representation of* B, $(H_{\tilde{\varphi}},\pi_{\tilde{\varphi}},\xi_{\tilde{\varphi}})$, *determined by* $\tilde{\varphi}$. *Then there exist an essentially selfadjoint operator* $h_{\tilde{\varphi}}$ *on* $H_{\tilde{\varphi}}$, *a *-subalgebra* $\mathcal{D}\subset\overline{\pi_{\tilde{\varphi}}(B)}^{\sigma}$ *and a derivation* δ_1 *of* $\mathcal{D}$ *which extends* $\delta^{\tilde{\varphi}}$ *such that:*

(i) $\pi_{\tilde{\varphi}}(\mathcal{D}(\delta))\subset\mathcal{D}$,

(ii) $\mathcal{D}(h_{\tilde{\varphi}})=\mathcal{D}\xi_{\tilde{\varphi}}$,

(iii) $\delta_1(a_{\tilde{\varphi}})\xi=\iota[h_{\tilde{\varphi}},a_{\tilde{\varphi}}]\xi$, $a_{\tilde{\varphi}}\in\mathcal{D}$, $\xi\in\mathcal{D}(h_{\tilde{\varphi}})$.

Moreover, we have $||\pi_{\tilde{\varphi}}[(1+\alpha\delta)(a)]||\geq||\pi_{\tilde{\varphi}}(a)||$, $\alpha\in\mathbb{R}$, $a\in\mathcal{D}(\delta)$.

PROOF. Using Lemmas 3.3-3.5 the proof of this lemma is a verbatim repetition of the proof of Lemma 2.4 with the following substitutions:

We replace:

$\hat{\pi}\in\hat{G}$	by	$\gamma\in\Gamma$,
$B(\hat{\pi})$	by	$\tilde{\rho}(A)\bar{\lambda}_{\gamma}$,
$\beta(\hat{\pi})$	by	P_{γ} ,
$M_{\tilde{\varphi}}(\hat{\pi})$	by	$\overline{\pi_{\tilde{\varphi}}(\tilde{\rho}(A)\bar{\lambda}_{\gamma})}^{\sigma}$.

THEOREM 3.7. *Let* (A,Γ,α) *be a* C*-*dynamical system with* Γ *discrete. Let* (B,β) *be the dual system. If* δ *is a symmetric, closed, densely defined derivation of* B *such that*

1) $\tilde{\rho}(A)\subset\mathcal{D}(\delta)$,

2) $\beta\delta\subset(\delta\bar{\otimes}\iota)\beta$,

then δ *is a generator.*

PROOF. By Lemma 3.3, $\Sigma\tilde{\rho}(A)\bar{\lambda}_{\gamma}\subset\mathcal{D}(\delta)$ and so δ possesses a dense set of analytic elements. On the other hand, by Lemma 3.6 we have

$$||(1+\alpha\delta)(a)||=\sup_{\varphi}||\pi_{\tilde{\varphi}}[(1+\alpha\delta)(a)]||\geq\sup_{\varphi}||\pi_{\tilde{\varphi}}(a)||=||a||$$

$a\in\mathcal{D}(\delta)$, $\alpha\in\mathbb{R}$, where the supremum is taken over all states φ of $\tilde{\rho}(A)$. Therefore δ satisfies the conditions A1), B2), C1) of [2,

Theorem 3.2.50] and so δ is a generator.

The preceding theorem is a significant improvement of [5, Theorem 3.5].

PROBLEM 3.8. Let (B,G,β) be a C^*-dynamical system with G compact. Let δ be a symmetric, closed, densely defined derivation of B. Suppose that

1) $\beta_g \delta \subset \delta \beta_g$ for all $g \in G$,

2) $B(\hat{\pi}_o) \subset \mathcal{D}(\delta)$.

Does it follow that δ is a generator ?

4. LIE ALGEBRAS OF DERIVATIONS

Let L be a finite dimensional real, Lie algebra, and let L be the corresponding connected, simply connected Lie group.

Let also B be C^*-algebra and $\mathcal{D} \subset B$ a $*$-subalgebra of B. We denote by $\mathrm{Der}^*_{\mathcal{D}}(B)$ the set of all symmetric derivations of B with common invariant domain $\mathcal{D}$.

A mapping $\Delta : L \to \mathrm{Der}^*_{\mathcal{D}}(B)$ is called a representation of L by symmetric derivations of B if:

(i) $\Delta(\ell_1+\ell_2)=\Delta(\ell_1)+\Delta(\ell_2)$ on $\mathcal{D}$ $\quad \ell_1,\ell_2 \in L$

(ii) $\Delta([\ell_1,\ell_2])=[\Delta(\ell_1),\Delta(\ell_2)]$ on $\mathcal{D}$ $\quad \ell_1,\ell_2 \in L$.

A representation Δ of L by symmetric derivations of B is called integrable if there exists a representation of $L, \tau : L \to \mathrm{Aut}(B)$ such that $\Delta(l)$ is a pregenerator of the one parameter group $\{\tau(\exp tl)\}_{t\in R}$ for each $\ell \in L$.

Let $\{\ell_1,\ldots,\ell_k\} \subset L$ be a set of Lie generators for L. Using Theorems 2.5, 3.7 and [3, Lemma 1] it is easy to prove the following two results.

THEOREM 4.1. *Let* (B,G,β) *be a* C^*-*dynamical system with* G *compact and let* Δ *be a representation of* L *by symmetric* $*$-*derivations of* B. *Denote by* $\mathcal{D}$ *the common invariant dense domain of the* $\Delta(\ell)$, $\ell \in L$. *Assume that:*

(i) $\sum_{\hat{\pi} \in \hat{G}} B(\hat{\pi}) \subset \mathcal{D}$,

(ii) $\beta_g \Delta(\ell_j) \subset \Delta(\ell_j)\beta_g$, $g \in G$, $j=1,\ldots,k$,

(iii) $\Delta(\ell_j)$ *is closable*, $j=1,\ldots,k$.

Then Δ *is integrable.*

THEOREM 4.2. *Let* (A,Γ,α) *be a* C^**-dynamical system with* Γ *discrete and* A *unital. Denote by* (B,β) *its dual system. Let also* Δ *be a representation of* L *by symmetric *-derivations of* B. *Denote by* $\mathcal{D}$ *the common invariant dense domain of the* $\Delta(\ell)$, $\ell \in L$. *Assume that:*

(i) $\tilde{\rho}(A) \subset \mathcal{D}$,

(ii) $\beta\delta \subset (\delta \,\overline{\otimes}\, \iota)\beta$,

(iii) $\Delta(\ell_j)$ *is closable*, $j=1,\ldots,k$.

Then Δ *is integrable.*

From [5, Theorem 2.5] and [3, Lemma 1] we also have.

THEOREM 4.3. *Let* (B,G,β) *be a* C^**-dynamical system with* G *compact and* B *unital. Asuume that* G *acts ergodically on* B. *Let* Δ *be a representation of* L *by symmetric *-derivations of* B. *Assume that* $\beta_g\Delta(\ell_j)$ $\Delta(\ell_j)\beta_g$ *for all* $g \in G$, $j=1,\ldots,k$. *Then* Δ *is integrable.*

REFERENCES

1. Bourbaki, N.: *Intégration*, Hermann, Paris, 1959.

2. Bratteli, O.; Robinson, D.W.: *Operator algebras and quantum statistical mechanics*. I, Springer Verlag, 1979.

3. Jørgensen, P.E.T.: Perturbation and analytic continuation of group representations, *Bull.Amer.Math.Soc.* 82(1976), 921-924.

4. Landstad, M.B.: Duality theory for covariant systems, *Trans. Amer.Math.Soc.* 248(1979), 223-267.

5. Peligrad, C.: Derivations of C^*-algebras which are invariant under an automorphisms group, in "*Topics in modern operator theory*",OT Series vol.2, Birkhäuser Verlag, 1981, pp.259-268.

6. Strătilă, Ş.; Zsidó, L.: *Operator algebras*, preprint, 1978.

7. Takesaki, M.: *Theory of operator algebras*. I, Springer Verlag, 1979.

8. Tomiyama, J.: On the projections of norm one in the direct product of operator algebras, *Tohôku Math.J.* 11(1959), 125-129.

9. Zsidó, L.: Reduction theory of W*-algebras, in "*Operator theory and operator algebras*" (Romanian), Ed.Academiei RSR, Bucharest, 1973, pp.131-262.

ACKNOWLEDGEMENTS

After this paper was presented at the VI-th Conference on "Operator Theory", Timişoara-Herculane, June 1981, we have received some preprints on similar problems which are listed in the supplementary bibliography below.

In [10] our Theorem 1.5 was given in the special case when G is commutative. Notice also that [10, Theorem 5.2] may be strengthened to solve our Problem 3.8 in the particular case when G is commutative. This fact will be considered in a subsequent paper.

SUPPLEMENTARY BIBLIOGRAPHY

10. Bratteli, O.; Jørgensen, P.E.T.: Unbounded derivations tangential to compact groups of automorphisms, *preprint*, 1981.

11. Goodman, F.; Jørgensen, P.E.T.: Unbounded derivations commuting with compact group actions, *preprint*, 1980/1981.

12. Ikunishi, A.: Derivations of C*-algebras commuting with compact actions, *preprint*, 1981.

C.Peligrad
Department of Mathematics,
INCREST,
Bdul Păcii 220, 79622 Bucharest,
Romania.

ON COMMUTATORS IN PROPERLY INFINITE W^*-ALGEBRAS

Sorin Popa

INTRODUCTION

Let M be an algebra. An element $x\in M$ is said to be a commutator if there are $a,b\in M$ such that $x=ab-ba$.

If M is unital Banach algebra then by the classical Wintner-Wielandt theorem no nonzero scalar is a commutator. Moreover if M is a matrix algebra then it is well known that $x\in M$ is a commutator if and only if trace$(x)=0$.

For the algebra of all bounded operators $\mathcal{B}(H)$ on a separable Hilbert space the problem of characterising the commutators was solved by A.Brown and C.Pearcy in [4]. Subsequently the method of [4] was extended in [5], [8] to yield the following more general result: let R be the maximal ideal of a properly infinite factor M, then $x\in M$ is a commutator iff $x\notin(\mathbb{C}\setminus\{0\})1_M+R$. Moreover, in case M is properly infinite with arbitrary center the same technique can be used to obtain good sufficient conditions for an element to be a commutator (see [9], [10]). More precisely these results are obtained by combining the Brown-Pearcy technique with reduction theory arguments.

However the general problem of characterizing all the commutators in a properly infinite W^*-algebra with rich center (i.e. infinite dimensional center) is still open (see [10]). The main difficulty in this problem seems not to lie in reduction-theory, but rather in the proof of a Wintner-Wielandt type result with estimates and of a Brown-Pearcy type result with estimates.

The aim of this paper is to give some partial results in this direction.

In Section 1 we give a necessary condition for an element to be a commutator (Theorem 1.1), based on estimates of the norm of a derivation whose range is close to the scalars.

In Section 2 we prove a sufficient condition for an element to be a commutator (Theorem 2.1) and we make some remarks concerning the problem.

Both conditions appearing in Theorems 1.1 and 2.1 are obtained in terms of the behavior of a generalized η function (of Brown-Pearcy) associated to an element of the algebra and which was studied in detail in [3].

For standard notations and basic results in operator algebras we refer the reader to [12].

§1

In what follows M will allways denote a properly infinite W*-algebra, Z its center and Ω the maximal ideal space of Z. By $[t]$ we denote the closed two-sided ideal in M generated by $t\in\Omega$ and by I_t we denote the maximal ideal in M with the property that $I_t \cap Z = t$ (cf.[11]).

Recall that for $x\in M$ the functions

$$\Omega \ni t \to d(x,[t]),$$
$$\Omega \ni t \to d(x,I_t),$$

are continuous (see [6], [3]). Also we mention that $||x|| = \sup_{t\in\Omega} d(x,[t])$ (cf.[6]).

It easily follows that

$$\Omega \ni t \to d(x,\mathbb{C}\cdot 1 + I_t)$$

is also continuous (see [3]). In fact the function $d(x,\mathbb{C}+I_t)$ is a generalized η function of A.Brown and C.Pearcy, as shown in [3] by C.Apostol and L.Zsidó.

An immediate consequence of the Wintner-Wielandt theorem is that if $x\in M$ is a commutator then its image in M/I_t must not be a nonzero scalar, or equivalently the functions $d(x,I_t)$ and $d(x,\mathbb{C}+I_t)$ must have the same zeros. But this condition is not sufficient if Z is infinite dimensional (see [10], Example 3.13).

In fact we will prove the following:

THEOREM 1.1. *If* $x\in M$ *is a commutator then*

$$\sup_{\substack{t\in\Omega \\ x\notin I_t}} d(x,I_t)\ln \frac{d(x,I_t)}{d(x,\mathbb{C}+I_t)} < \infty.$$

First we need a preliminary result.

LEMMA 1.2. *Let* A *be a Banach algebra with unit* $1\in A, ||1||=1$ *and* $\delta:A\to A$ *a derivation. If*

$$\inf\{||1-\delta(a)|| \mid a\in A,\ ||a||=1\}=\varepsilon,$$

then

$$||\delta||\geq \ln \frac{1}{\varepsilon}.$$

Proof. Let $a\in A$, $||a||=1$, $||1-\delta(a)||=\varepsilon'\geq\varepsilon$, and denote by $x=1-\delta(a)$. Then for arbitrary $k\geq 1$ we have:

$$\delta(a^k)=(1-x)a^{k-1}+a(1-x)a^{k-2}+\ldots+a^{k-1}(1-x)=$$
$$=ka^{k-1}-(xa^{k-1}+axa^{k-2}+\ldots+a^{k-1}x),$$

so that

$$||\delta||\cdot||a^k||\geq k||a^{k-1}||-k\varepsilon'.$$

For each $k\geq 1$ multiply the above inequality by $\frac{1}{k!}||\delta||^{k-1}$ and then add all the inequalities for $k\geq 1$ to get:

$$\sum_{k\geq 1}\frac{1}{k!}||\delta||^k||a^k||\geq$$
$$\geq \sum_{k\geq 0}\frac{1}{k!}||\delta||^k||a^k||-\varepsilon'\sum_{k\geq 0}\frac{1}{k!}||\delta||^k,$$

or equivalently:

$$0\geq 1-\varepsilon'\exp(||\delta||)$$

which gives

$$||\delta||\geq\ln\frac{1}{\varepsilon'}.$$

REMARK 1.3. The proof is also valid in the more general case when $A\subset B$ are normed algebras with the same unit 1, $||1||=1$ and $\delta:A\to B$ is a derivation.

As an immediate consequence of the lemma we get the following:

COROLLARY 1.4. *Let* A *be a Banach algebra with unit* $1\in A, ||1||=1$. *If* $a,b\in A$, $||a||=1$ *and*

$$||(ab-ba)-1||\leq\varepsilon$$

then

$$||b||\geq\frac{1}{2}\ln\frac{1}{\varepsilon}.$$

REMARK 1.5. In [7], Problem 183, it is shown that if a_n, $b_n \in A$ and $a_n b_n - b_n a_n$ tends to 1, then $||a_n||+||b_n||$ tends to infinity. The Corollary gives an estimate of how fast $||a_n||+||b_n||$ increase.

PROOF OF THEOREM 1.1. Let $x=ab-ba$ and for arbitrary $y \in M$ denote by y_t the image of y in M/I_t. Then $x_t = a_t b_t - b_t a_t$ and $||a_t|| \le ||a||$, $||b_t|| \le ||b||$, for all $t \in \Omega$. Let $\lambda_t \in \mathbb{C}$ be such that

$$||x_t - \lambda_t|| = d(x, \mathbb{C}+I_t).$$

Applying the corollary for

$$||(a_t b_t - b_t a_t) - \lambda_t|| = d(x, \mathbb{C}+I_t)$$

we obtain that

$$\frac{1}{|\lambda_t|}||a_t|| \cdot ||b_t|| \ge \frac{1}{2} \ln \frac{|\lambda_t|}{d(x, \mathbb{C}+I_t)},$$

so that

$$(*) \qquad 2||a|| \cdot ||b|| \ge |\lambda_t| \ln \frac{|\lambda_t|}{d(x, \mathbb{C}+I_t)}.$$

If $t \in \Omega$ is such that $\frac{d(x, I_t)}{d(x, \mathbb{C}+I_t)} < 2$ then

$$d(x, I_t) \ln \frac{d(x, I_t)}{d(x, \mathbb{C}+I_t)} \le 2||x||.$$

If $\frac{d(x, I_t)}{d(x, \mathbb{C}+I_t)} \ge 2$ then

$$||x_t|| = d(x, I_t) \ge 2d(x, \mathbb{C}+I_t) =$$
$$= 2||x_t - \lambda_t|| \ge 2||x_t|| - 2|\lambda_t|,$$

so that

$$||x_t|| = d(x, I_t) \ge |\lambda_t| \ge \frac{1}{2} d(x, I_t)$$

and the statement of the theorem follows by (*).

§2

Sufficient conditions for an element in M to be a commutator have been proved by H.Halpern in [9]: it is shown that if

$$\sup_{\substack{t \in \Omega \\ x \notin I_t}} \frac{d(x, I_t)}{d(x, \mathbb{C}+I_t)} < \infty$$

then x is a commutator. The proof uses the following technique: the algebra M is cut in a sequence of pieces by projections of

$\mathcal{Z}$ and on each piece, applying the Brown-Pearcy technique, the element x has by similarity a suitable standard form which is a commutator. When the pieces are put together again the elements which give locally x as a commutator must have uniform bounded norm and this is why one needs the above condition.

In an earlier paper [10] H.Halpern shows that in fact

$$\sup_{\substack{t\in\Omega \\ x\notin I_t}} \frac{d(x,I_t)}{d(x,\mathbb{C}+I_t)}<\infty$$

if and only if x is similar to a selfadjoint commutator (i.e.there are $a,b\in M$ such that $a=a^*$ and x is similar to ab-ba). This is a generalisation of a theorem of J.Anderson ([1]) on selfadjoint commutators in $\mathcal{B}(\mathcal{H})$.

In the same paper H.Halpern points out that there are commutators which do not satisfy the above condition (see Example 3.15, [10]): this is because the condition is invariant under multiplication of x by elements in $\mathcal{Z}$, while every $x\in M$ for which $d(x,I_t)=0$ implies $d(x,\mathbb{C}+I_t)=0$ can be multiplied by an element $z\in\mathcal{Z}$, $s(z)=1$, such that zx is a commutator. We prove now a more precise result in this direction:

THEOREM 2.1. *If* $x\in M$ *is such that*

$$\sup_{\substack{t\in\Omega \\ x\in I_t}} \frac{d(x,[t])^{3/2}}{d(x,\mathbb{C}+I_t)} < \infty$$

then x *is a commutator.*

For the reduction-theory part of the proof of this theorem we need the next two lemmas.

LEMMA 2.2. *If* $p\in M$ *is a projection and* $p\notin I_t$ *then there exists a central projection* $z\in\mathcal{Z}$ *such that* $zp\sim z$ *and* $z(t)\neq 0$ (i.e. $z\notin t$).

PROOF. Denote by $I=\mathrm{span}(MpM)$, the two sided ideal generated by p in M.Since I_t is a maximal ideal in M and $I_t\subsetneq I_t+I$, it follows that the ideal I_t+I must equal M. In particular there exist $z_1\in I_t$ and $x_1,\ldots,x_m$, $y_1,\ldots,y_m\in M$ such that

$$z_1+\sum_{i=1}^{m} x_i p y_i=1$$

If we regard $1-z_1\in \mathcal{Z}=C(\Omega)$ as a continuous complex function on Ω, then $(1-z_1)(t)=1$ so that there exists an open closed neighborhood V of t in Ω such that:

$$V\subset \{t\in\Omega \mid \tfrac{3}{2}\geq |(1-z_1)(t)|\geq\tfrac{1}{2}\}.$$

If z denotes the projection in $\mathcal{Z}$ corresponding to the characteristic function of V, then multiplying the equality $\sum_{i=1}^{m} x_i p y_i = 1-z_1$ by a suitable element of $\mathcal{Z}$, we may suppose that $p\leq z$ and that $\sum_{i=1}^{m} x_i p y_i=z$.

Let $z_o\in z\mathcal{Z}$ be a central projection such that $z_o p$ is a finite projection. Then

$$z_o=\sum_{i=1}^{m} x_i (z_o p) y_i.$$

Since $\ell(x_i(z_o p)y_i)\precsim z_o p$, $i=1,\dots,m$ (for $y\in M$, $\ell(y)$ is the left support of y in M), it follows that $\ell(x_i(z_o p)y_i)$ is finite and

$$z_o=\ell\Big(\sum_{i=1}^{m} x_i (z_o p) y_i\Big)\leq\bigvee_{i=1}^{m} \ell(x_i(z_o p)y_i)$$

is also finite. But this is contradictory unless $z_o=0$. This shows that p is a properly infinite projection and by $\ell(x_i p y_i)\precsim p$ it follows that

$$z=\bigvee_{i=1}^{m}\ell(x_i p y_i)\precsim p.$$

LEMMA 2.3. *If* $x\in M$, $t\in\Omega$, $d(x,\mathbb{C}+I_t)=\alpha>0$, *then there exists a projection* p *in* M *such that*:

1) $2\alpha p\geq |(1-p)xp|\geq\frac{1}{2}\alpha p$;
2) $p\notin I_t$, $1-p\notin I_t$;
3) $1-(p+\ell((1-p)xp))\notin I_t$;

(*here* $|y|$ *is the module of* y *and* $\ell(y)$ *is the left support of* $y\in M$).

PROOF. By [3] we can find a projection $q_t\in M/I_t$ such that if x_t is the image of x in $M|I_t$ then

$$||(1-q_t)x_t q_t||\geq\tfrac{2}{3}\alpha.$$

Take $q\in P(M)$ to be a preimage for q_t and denote by p_o the spectral projection of $|(1-q)xq|$ corresponding to the interval $[\frac{1}{2}\alpha,\infty)$. Then p_o satisfies $p_o\notin I_t$, $1-p_o\notin I_t$ (since $1-p_o\geq 1-q\notin I_t$) and we have

$$|(1-q)xq|p_o \geq \frac{1}{2}\alpha p_o.$$

We also have:

$$p_o x^*(1-p_o)xp_o = p_o q x^*(1-q)xqp_o +$$
$$+p_o x^*(q-p_o)xp_o \geq p_o|(1-q)xq|^2 p_o \geq \frac{1}{4}\alpha^2 p_o.$$

It follows that:

$$|(1-p_o)xp_o| \geq \frac{1}{2}\alpha p_o$$

and by the above computations any projection $p \leq p_o$ satisfies this condition.

Take $\lambda \in \mathbb{C}$ and $y = yp_o \in I_t$ such that

$$||(x-y-\lambda)p_o|| \leq \frac{3}{2}\alpha.$$

If $e \in I_t$ is a spectral projection of $|y|$, such that $||y(1-e)|| \leq \frac{1}{2}\alpha$, then $e \leq p_o$ and $p_1 = p_o - e$ satisfies

$$||(1-p_1)xp_1|| = ||(1-p_1)(x-\lambda)p_1|| \leq$$
$$\leq ||(1-p_1)(x-y-\lambda)p_1|| + ||(1-p_1)yp_1|| \leq$$
$$\leq ||(x-y-\lambda)p_o|| + ||y(1-e)|| \leq \frac{3}{2}\alpha + \frac{1}{2}\alpha = 2\alpha.$$

Thus p_1 satisfies 1) and 2).

Moreover, if $p \leq p_1$ is an arbitrary projection, then

$$||(1-p)xp|| \leq ||(1-p)(x-y-\lambda)p|| +$$
$$+||(1-p)yp|| \leq ||(x-y-\lambda)p|| + ||yp|| \leq$$
$$\leq ||(x-y-\lambda)p_o|| + ||yp_1|| \leq 2\alpha.$$

Taking $p \leq p_1$ such that $p \sim p_1 \sim p_1 - p$, condition 3) also follows.

For the local part in the proof of Theorem 2.1 we use Apostol's proof of the Brown-Pearcy theorem, because it also gives some norm estimates which we need.

LEMMA 2.4. (see [2], Lemma 2.3). *Let* x *be a two by two matrix over* M, $x = \begin{bmatrix} x_1 & \alpha v^* \\ x_2 & x_3 \end{bmatrix}$, *with* $x_1, x_2, x_3, v \in M, \alpha > 0$ *and* v *is an isometry with* $1 - vv^* \sim 1$.

Then there are two by two matrices over M, a *and* b, *such that*

$$x = \alpha(ab-ba),$$
$$||a|| \leq 2 + 3\alpha^{-1}(||x_1|| + ||x_2|| + ||x_3||),$$
$$||b|| \leq 1 + 10\alpha^{-1}(||x_1|| + ||x_2|| + ||x_3||) +$$
$$+48\alpha^{-2}(||x_1|| + ||x_2|| + ||x_3||)^2.$$

The proof is exactly the same as in the particular case $M = \mathcal{B}(H)$ treated by C.Apostol in [2]. We give here only a sketch of the proof.

Let $u\in M$ be an isometry such that $uu^*+vv^*=1$. For $y\in M$ denote by

$$S_v(y)=\sum_{k=0}^{\infty} v^k((1-vv^*)y(1-vv^*))v^{*k}\in M$$

$$S_u(y)=\sum_{k=0}^{\infty} u^k((1-uu^*)y(1-uu^*))u^{*k}\in M.$$

Using this notations, let

$$y_1=vS_v(x_1+x_3)-uu^*(x_1+x_3)v,$$
$$y_2=uS_u(x_1+x_3)-vv^*(x_1+x_3)u,$$
$$r=\frac{1}{2}(u^*y_1-y_1u^*)+vv^*x_3-uu^*x_1,$$
$$y=r(y_1+2y_2)+y_1r.$$

If we define:

$$a=\begin{bmatrix} \frac{1}{2}u^* & 0 \\ \alpha^{-1}r & \frac{1}{2}u^*-v^* \end{bmatrix}$$

$$b=\begin{bmatrix} \alpha^{-1}(y_1+2y_2) & 1 \\ \alpha^{-1}\sum_{n=0}^{\infty}(-1)^{n+1}2^{-n}v^{n+1}(x_2-\alpha^{-1}y)u^{*n} & -\alpha^{-1}y_1 \end{bmatrix},$$

then it is not hard to verify that

$$x=\alpha(ab-ba).$$

Moreover we have the following estimates:

$$||y_1||\le 2||x_1+x_3||,$$
$$||y_2||\le 2||x_1+x_3||,$$
$$||r||\le||y_1||+||x_3||+||x_1||\le 3(||x_1||+||x_3||),$$
$$||y||\le||r||(||y_1||+2||y_2||)+||y_1||\cdot||r||=$$
$$=2||r||(||y_1||+||y_2||)\le$$
$$\le 6(||x_1||+||x_3||)(4||x_1+x_3||)\le$$
$$\le 24(||x_1||+||x_3||)^2.$$

It follows that:

$$||a||\le 2+\alpha^{-1}||r||\le 2+\alpha^{-1}(||x_1||+||x_3||),$$
$$||b||\le 1+2\alpha^{-1}(||y_1||+||y_2||)+$$
$$+2\alpha^{-1}||x_2-\alpha^{-1}y||\le 1+8\alpha^{-1}||x_1+x_3||+$$
$$+2\alpha^{-1}||x_2||+48\alpha^{-2}(||x_1||+||x_3||)^2.$$

LEMMA 2.5. *Let* x *be a two by two matrix over* M,

$x=\begin{bmatrix} x_1 & \alpha v \\ x_2 & x_3 \end{bmatrix}$, *with* $x_1, x_2, x_3, v\in M$, $\alpha>0$ *and* v *is an isometry with* $1-vv^*\sim 1$. *Then there are* a,b *satisfying the same estimates as in Lemma* 2.4, *such that*:

$$x=\alpha(ab-ba).$$

PROOF. Let $s=\begin{bmatrix} 0 & 1 \\ 1 & 0 \end{bmatrix}$. Then $sx^*s^{-1}=\begin{bmatrix} x_1^* & \alpha v^* \\ x_2^* & x_1^* \end{bmatrix}$ and by Lemma 2.4 there exist a', b' such that $sx^*s^{-1}=\alpha(a'b'-b'a')$, a' and b' satisfying the estimates of Lemma 2.4. If we denote

$$a=-sa'^*s^{-1}$$
$$b=sb'^*s^{-1}$$

then $x=\alpha(ab-ba)$ and $||a||=||a'||$, $||b||=||b'||$ satisfy the required estimates.

PROOF OF THEOREM 2.1. Suppose $x\in M$, $||x||\le 1$, satisfies the hypothesis of Theorem 2.1.

Define recursively open closed disjoint sets $\Omega_n\subset\Omega$, such that:

$$\Omega_n\subset\{t\in\Omega\,|\,\frac{1}{2^{n+1}}\le d(x,\mathbb{C}+I_t)\le\frac{1}{2^n}\}$$
$$\Omega_n\supset\{t\in\Omega\,|\,\frac{1}{2^{n+1}}<d(x,\mathbb{C}+I_t)<\frac{1}{2^n}\}$$
$$\bigcup_{k=0}^{n}\Omega_k\supset\{t\in\Omega\,|\,\frac{1}{2^{n+1}}<d(x,\mathbb{C}+I_t)\le 1\}.$$

Denote by z_n the projections in Z corresponding to the characteristic functions of Ω_n and let $z=\sum_{n=0}^{\infty} z_n$. It is known by [5], [8] that $(1-z)x$ is a commutator so that in what follows we may suppose $z=1$.

Applying Lemma 2.2 and Lemma 2.3 for each $t\in\Omega_n$ and by the compacity of Ω_n one can find:

1) mutual orthogonal projections $\{z_n^k\}_{1\le k\le r_n}$ in Z with $\sum_k z_n^k=z_n$;

2) projections $\{p_n^k\}_{1\le k\le r_n}$ in M such that

a) $p_n^k\le z_n^k$, $p_n^k\sim z_n^k\sim z_n^k-p_n^k$

b) $2\cdot\frac{1}{2^n}p_n^k \geq |(1-p_n^k)xp_n^k| \geq \frac{1}{2}\cdot\frac{1}{2^n}p_n^k$

c) $(z_n^k-p_n^k)-\ell((z_n^k-p_n^k)xp_n^k)\sim z_n^k$.

Thus, by Lemma 2.3, if $p_n=\sum_k p_n^k$ and $s_n=(z_n-p_n)+2^n|(z_n-p_n)xp_n|\varepsilon z_nM$, then s_n is invertible and

$$||s_n||\leq 2,\ ||s_n^{-1}||\leq 2.$$

Denote by $v_n=2^n(z_n-p_n)(s_n\ x\ s_n^{-1})p_n$ and remark that v_n is a partial isometry. Moreover

$$v_n^*v_n=p_n$$
$$(z_n-p_n)-v_nv_n^*\sim z_n.$$

Thus if we identify z_nM with the two by two matrix over z_nM, via some equivalence of the projections p_n, z_n-p_n, z_n, then $x_n'=$ $=s_n(z_nx)s_n^{-1}$ has the form:

$$x_n'=\begin{pmatrix} x_1^n & \frac{1}{2^n}v_n \\ x_2^n & x_3^n \end{pmatrix}$$

Using the hypothesis of the theorem, if we denote by K a convenient constant, we get for i=1,2,3, n≥0:

$$||x_i^n||\leq||s_n(z_nx)s_n^{-1}||\leq 4||z_nx||=$$
$$=4\sup_{t\varepsilon\Omega_n} d(x,[t])\leq$$
$$\leq k\sup_{t\varepsilon\Omega_n} d(x,\mathbb{C}+I_t)^{2/3}\leq k\left(\frac{1}{2^n}\right)^{2/3}.$$

By Lemma 2.5 we get $a_n',b_n'\varepsilon z_nM$ such that

$$x_n'=\frac{1}{2^n}(a_n'b_n'-b_n'a_n'),$$
$$||a_n'||\leq K\left(\frac{1}{2^n}\right)^{2/3}\left(\frac{1}{2^n}\right)^{-1},$$
$$||b_n'||\leq K\left(\frac{1}{2^n}\right)^{4/3}\left(\frac{1}{2^n}\right)^{-2}.$$

Thus, if we take

$$a'=\sum_{n=0}^{\infty}\left(\frac{1}{2^n}\right)^{1/3}a_n'z_n,$$
$$b'=\sum_{n=0}^{\infty}\left(\frac{1}{2^n}\right)^{2/3}b_n'z_n,$$

then a',b' have bounded norm in M and

$$x'=\sum_{n=0}^{\infty}x_n'z_n=a'b'-b'a'.$$

Since $s=\sum_{n=0}^{\infty} s_n z_n$ is a bounded invertible element in M, it follows that

$$x=s^{-1}x's=ab-ba$$

with

$$a=s^{-1}a's,\ b=s^{-1}b's .$$

Since $I_t=[t]+R$, where $R=\bigcap_{t\in\Omega} I_t$ is the strong radical of M (see [3]), Theorem 2.1 easily yields:

COROLLARY 2.6. *If* $x\in M$ *is such that*

$$\sup_{t\in\Omega,\, x\notin I_t} \frac{d(x,I_t)^{3/2}}{d(x,\mathbb{C}+I_t)}<\infty$$

then there exists an element $y\in R$ *such that* $x-y$ *is a commutator.*

REMARK 2.7. In fact we strongly believe that the set of commutators elements in M is stable under perturbations by elements in the radical R of M. Unfortunately we could not prove that the weaker condition

$$\sup_{t\in\Omega,\, x\notin I_t} \frac{d(x,I_t)^{3/2}}{d(x,\mathbb{C}+I_t)}<\infty$$

is sufficient for x to be a commutator.

EXAMPLE 2.8. Let $M=\ell^{\infty}(\mathbb{N},\mathcal{B}(\mathcal{H}))$ and suppose $\mathcal{H}$ is separable infinite dimensional. For $T\in\mathcal{B}(\mathcal{H})$ denote by $||T||_e$ the essential norm of T (i.e. the distance of T to the compacts) and by $\eta(T)$ the usual η function of Brown-Pearcy (i.e. $\eta(T)$ is the distance of T to the thin operators).

Theorem 1.1 asserts that if $(T_n)_n\in\ell^{\infty}(\mathbb{N},\ \mathcal{B}(\mathcal{H}))$ is a commutator then

$$\sup_{\substack{n \\ T_n\in\mathcal{K}(\mathcal{H})}} ||T_n||_e \ln\frac{||T_n||_e}{\eta(T_n)}<\infty.$$

($\mathcal{K}(\mathcal{H})\subset\mathcal{B}(\mathcal{H})$ is the set of compact operators).

For instance if E_n are infinite dimensional projections of infinite dimensional corange then

$$(\tfrac{1}{n}I_{\mathcal{H}}+\tfrac{1}{n}\exp(-n^{1+\varepsilon})E_n)_n\in\ell^{\infty}(\mathbb{N},\mathcal{B}(\mathcal{H}))$$

is not a commutator, for $\varepsilon>0$.

Theorem 2.1 asserts that if $(T_n)_n \in \ell^\infty(\mathbb{N}, \mathcal{B}(H))$ satisfies

$$\sup_{\substack{n \\ T_n \notin \mathcal{K}(H)}} ||T_n||^{1/2} \frac{||T_n||}{\eta(T_n)} < \infty$$

then $(T_n)_n$ is a commutator.

For instance, with the above notations, the element $(\frac{1}{n} I_H + \frac{1}{n^{3/2}} E_n)_n \in \ell^\infty(\mathbb{N}, \mathcal{B}(H))$ is a commutator.

However we were not able to prove that $(\frac{1}{n} I_H + \frac{1}{n^{3/2}} E_n + K)_n$ is a commutator, for $K \in \mathcal{K}(H)$ an arbitrary compact operator.

REMARK 2.9. To solve the general problem the reduction theory part does not seem to be difficult. The main difficulties are already present if $M = \ell^\infty(\mathbb{N}, \mathcal{B}(H))$ so that this is the important case to be clarified. To give a complete answer for the $\ell^\infty(\mathbb{N}, \mathcal{B}(H))$-case one needs to solve two kinds of problems:

1) If $S, T \in \mathcal{B}(H)$, $||(ST-TS)-I|| = \varepsilon$ is small then find good estimates for $||S|| \cdot ||T||$. For instance, if the condition

$$\sup_{\substack{n \\ T_n \notin \mathcal{K}(H)}} ||T_n||_e^{1/2} \frac{||T_n||_e}{\eta(T_n)} < \infty,$$

mentioned in Remark 2.7, would be necessary and sufficient, then for the necessary part we need to have:

$$||T|| \cdot ||S|| \geq c \cdot \varepsilon^{-2},$$

where c is a universal constant.

2) If $X \in \mathcal{B}(H)$ is a commutator, then find $S, T \in \mathcal{B}(H)$, such that

$$X = ST - TS$$

and $||S|| \cdot ||T||$ as small as possible.

If this is done by similarity, then one needs good estimates for the norm of the similarity and for the norm of the elements which give X as a commutator in the given standard form.

REFERENCES

1. Anderson, J.: *Derivations, commutators and the essential numerical range*, Thesis, Indiana University, 1971.

2. Apostol, C.: Commutators on Hilbert space, *Revue Roumaine Math.Pures Appl.* 18(1973), 1013-1024.

3. Apostol, C.; Zsidó, L.: Ideals in W*-algebras and the function η of A.Brown and C.Pearcy, *Revue Roumaine Math.Pures Appl.* 18(1973), 1151-1170.

4. Brown, A.; Pearcy, C.: Structure of commutators of operators, *Ann.of Math.* 82(1965), 112-127.

5. Brown, A.; Pearcy, C.; Topping D.: Commutators and the strong radical, *Duke Math.J.* 35(1968), 853-860.

6. Glimm, J.: A Stone-Weierstrass theorem for C*-algebras, *Ann. of Math.* 72(1960), 216-244.

7. Halmos, P.: *A Hilbert space problem book*, von Nostrand, Princeton, 1967.

8. Halpern, H.: Commutators in properly infinite von Neumann algebras, *Trans.Amer.Math.Soc.* 139(1969), 55-74.

9. Halpern, H.: Commutators modulo the center in a properly infinite von Neumann algebra, *Trans.Amer.Math.Soc.* 150(1970), 55-68.

10. Halpern, H.: Essential central range and selfadjoint commutators in properly infinite von Neumann algebras, *Trans. Amer.Math.Soc.* 228(1977), 117-146.

11. Misonou, Y.: On a weakly central operator algebra, *Tôhoku Math.J.* 4(1952), 194-202.

12. Strătilă, Ș.; Zsidó, L.: *Lectures on von Neumann algebras*, Abacus Press, 1979.

Sorin Popa
Department of Mathematics,
INCREST,
Bdul Păcii 220, 79622 Bucharest,
Romania.

A FUNCTIONAL MODEL FOR THE UNITARY DILATION OF A POSITIVE DEFINITE MAP

Ion Suciu

1. Let E be a separable Hilbert space and let $\Gamma:\mathbb{Z}\to L(E)$ be a positive definite map on the additive group $\mathbb{Z}$ of the integers whose values are bounded operators on E. Suppose $\Gamma(0)=I$, the identity operator on E. According to Naimark dilation theorem there exist a Hilbert space H containing E as a subspace and a unitary operator W on H such that

(1.1) $\quad \Gamma(n)a=P_E W^n a, \qquad a\in E,\ n\in\mathbb{Z}.$

Under the minimality condition

(1.2) $$H=\bigvee_{-\infty}^{\infty} W^n E$$

the pair $\{E, W\}$ is uniquely determined, up to a unitarity which conserves the embedding of the initial space E into H.

Let E be the spectral measure of the operator W^* and F be the $L(E)$-valued semi-spectral measure on the unit circle $\mathbb{C}$ obtained by compressing E to E, i.e. for a Borel set σ of $\mathbb{C}$

(1.3) $\quad F(\sigma)=P_E E(\sigma)\,|\,E.$

The semi-spectral measure F represents Γ in the sense that

(1.4) $$\Gamma(n)=\int_0^{2\pi} e^{-int}dF(t), \qquad n\in\mathbb{Z},$$

and it is uniquely determined by Γ.

Let us define

(1.5) $$N_\Gamma(\lambda)=\int_0^{2\pi}\frac{e^{it}+\lambda}{e^{it}-\lambda}\,dF(t), \qquad |\lambda|<1.$$

Then for each λ in the unit disc $\mathbb{D}$, $N_\Gamma(\lambda)$ is a bounded operator on E and Re $N_\Gamma(\lambda)\geq 0$. The function $\lambda\to N_\Gamma(\lambda)$ is an $L(E)$-valued analytic function on $\mathbb{D}$. It results that the function $\Theta_\Gamma(\lambda)$ defined by

(1.6) $\Theta_\Gamma(\lambda)=[I-N_\Gamma(\lambda)][I+N_\Gamma(\lambda)]^{-1}$

is an $L(E)$-valued contractive analytic function on $\mathbb{D}$. Note that $N_\Gamma(\lambda)$ and consequently $\Theta_\Gamma(\lambda)$, depends on Γ only and not on particulary chosen model for the pair $\{H, W\}$.

Let $\Theta_\Gamma(e^{it})$ be the boundary Fatou limit of $\Theta_\Gamma(\lambda)$ (i.e. $\Theta_\Gamma(\lambda)$ tends strongly a.e. to $\Theta_\Gamma(e^{it})$ whenever λ tends nontangentially to e^{it}) and

$$\Delta_\Gamma(e^{it})=[I-\Theta_\Gamma(e^{it})^*\Theta_\Gamma(e^{it})]^{1/2}.$$

Then $\Theta_\Gamma(e^{it})$ and $\Delta_\Gamma(e^{it})$ are measurable function on the unit circle whose values are contractions on E. Let us denote by Θ_Γ and Δ_Γ the contractions on $L^2(E)$ defined by pointwise multiplications by $\Theta_\Gamma(e^{it})$ and $\Delta_\Gamma(e^{it})$ respectively. Note that $\Theta_\Gamma L_+^2(E)\subset L_+^2(E)$ where $L_+^2(E)$ is the subspace of $L^2(E)$ consisting from all the functions in $L^2(E)$ whose negative Fourier coefficients are zero. The subspace $L_+^2(E)$ can be identified with the space $H^2(E)$ of all the functions which are analytic on $\mathbb{D}$, take values in E and have boundary limits in $L^2(E)$. In this identification $(\Theta_\Gamma f)(\lambda)=\Theta_\Gamma(\lambda)f(\lambda)$, $\lambda\in\mathbb{D}$, $f\in H^2(E)$.

The aim of this note is to prove the following:

THEOREM. *Let* E, Γ, Θ_Γ, Δ_Γ *be as above. Let* $\hat{H}$ *be the Hilbert space*

$$\hat{H}=[L_+^2(E)\oplus\overline{\Delta_\Gamma L^2(E)}]\ominus[\Theta_\Gamma f\oplus\Delta_\Gamma f,\ f\in L_+^2(E)],$$

$\hat{W}$ *be the operator defined on* $\hat{H}$ *by*

$$\hat{W}(u\oplus v)=e^{it}[u-\Theta_\Gamma u(0)-u(0)]\oplus e^{-it}[v-\Delta_\Gamma u(0)],\ u\oplus v\in\hat{H}$$

and V *be the operator from* E *into* $\hat{H}$ *defined by*

$$Va=e^{-it}\Theta_\Gamma a\oplus e^{-it}\Delta_\Gamma a, \qquad a\in E.$$

Then $\hat{W}$ *is a unitary operator on* $\hat{H}$, V *is an isometry from* E *into* $\hat{H}$ *and*

$$\Gamma(n)=V^*\hat{W}^nV, \qquad n\in\mathbb{Z}.$$

Moreover

$$\hat{H}=\bigvee_{-\infty}^{\infty}\hat{W}^nVE,$$

and for any $u\oplus v$ *in* $\hat{H}$ *we have*

$$V^*(u\oplus v)=\frac{1}{2\pi}\int_0^{2\pi} e^{it}[\Theta_\Gamma(e^{it})^*u(e^{it})+\Delta_\Gamma(e^{it})v(e^{it})]dt.$$

The ideas of the proof of the theorem are mainly contained in the paper [1] of M.S.Brodskiĭ. In fact the proof will be obtained as a consequence of some lemmas which will explicitate the elements from the Nagy-Foiaş model for contractions used implicitely in the above quoted paper of M.S.Brodskiĭ.

2. Let W, acting on H, be a given identification for the minimal unitary dilation of Γ. So we have $E\subset H$ and (1.1) and (1.2) hold. Let us denote

(2.1) $\quad T=W^*[I_H-P_E]$

and, as usually, $D_T=[I-T^*T]^{1/2}$, $D_{T^*}=[I-TT^*]^{1/2}$, $\mathcal{D}_T=\overline{D_T H}$, $\mathcal{D}_{T^*}=\overline{D_{T^*}H}$.

LEMMA 1. (i) T *is a completely non unitary contraction on* H.

(ii) $\quad D_T=P_E$, $D_{T^*}=W^*P_E W$, $\mathcal{D}_T=E$, $\mathcal{D}_{T^*}=W^*E$.

(iii) $\quad TD_T=0$, $T^*D_{T^*}=0$.

PROOF. Let H_o be a subspace of H which reduces T to a unitary operator. For any $h\in H_o$ we have

$$||h||=||Th||=||W^*[I-P_E]h||=||(I-P_E)h||$$

which implies $h=(I-P_E)h$ and consequently $H_o\perp E$. Since for $h\in H_o$ we also have

$$||Wh||=||h||=||T^*h||=||(I-P_E)Wh||,$$

it results

$$Wh=[I-P_E]Wh=T^*h\in H_o$$
$$W^*h=W^*[I-P_E]h=Th\in H_o.$$

This means that H_o reduces W and from the minimality condition (1.2) we obtain $H_o=\{0\}$, i.e. T is completely non unitary.

Since

$$I-T^*T=I-[I-P_E]WW^*[I-P_E]=P_E$$
$$I-TT^*=I-W^*[I-P_E]W=W^*P_E W$$

the points (ii) and (iii) of the lemma follow.

Recall (cf.[2], Ch.VI,Sec.1) that for a contraction T on H its characteristic function is the contractive analytic function $\{\mathcal{D}_T, \mathcal{D}_{T^*}, \Theta_T(\lambda)\}$ defined by

(2.2) $\quad \Theta_T(\lambda)=[-T+\lambda D_{T^*}[I-\lambda T^*]^{-1}D_T]|\mathcal{D}_T.$

Having in mind the form of D_T, D_{T^*}, $\mathcal{D}_T$ and $\mathcal{D}_{T^*}$ given in Lemma 1 it is clear that, in our case, $\Theta_T(\lambda)$ is the operator from E into W^*E given by

(2.3) $\Theta_T(\lambda)a=\lambda W^*P_E W[I-\lambda T^*]^{-1}a, \qquad a\varepsilon E.$

LEMMA 2. *We have*

$$\Theta_\Gamma(\lambda)=-W\Theta_T(\lambda).$$

PROOF. From (1.5), (1.4) and (1.1) it results

$$N_\Gamma(\lambda)a=P_E[W^*+\lambda][W^*-\lambda]^{-1}a, \qquad a\varepsilon E.$$

Then for $b\varepsilon E$ we obtain

$$[I_E-N_\Gamma(\lambda)]b=[I_H-P_E[W^*+\lambda][W^*-\lambda]^{-1}]b=$$
$$=P_E[I_H-[W^*+\lambda][W^*-\lambda]^{-1}]b=P_E[W^*-\lambda-W^*-\lambda][W^*-\lambda]^{-1}b=$$
$$=-2\lambda P_E[W^*-\lambda]^{-1}b.$$

Let $b=[I_E+N_\Gamma(\lambda)]^{-1}a$, $a\varepsilon E$. Then

$$a=[I_E+N_\Gamma(\lambda)]b=[I_H+P_E[W^*+\lambda][W^*-\lambda]^{-1}]b=$$
$$=[W^*-\lambda+P_E W^*+\lambda P_E][W^*-\lambda]^{-1}b=$$
$$=[[I_H+P_E]W^*-\lambda[I-P_E]][W^*-\lambda]^{-1}b=$$
$$=[I_H+P_E][W^*-\lambda[I_H+P_E]^{-1}[I-P_E]][W^*-\lambda]^{-1}b=$$
$$=[[I_H+P_E]W^*-\lambda[I-P_E]][W^*-\lambda]^{-1}b=$$
$$=[I_H+P_E][I_H-\lambda[I_H-P_E]W]W^*[W^*-\lambda]^{-1}b=$$
$$=[I_H+P_E][I-\lambda T^*]W^*[W^*-\lambda]^{-1}b.$$

It results

$$b=[W^*-\lambda]W[I-\lambda T^*]^{-1}[I_H+P_E]^{-1}a=$$
$$=\frac{1}{2}[W^*-\lambda]W[I_H-\lambda T^*]^{-1}a.$$

Hence for any $a\varepsilon E$ we obtain

$$\Theta_\Gamma(\lambda)a=[I_E-N_\Gamma(\lambda)][I_E+N_\Gamma(\lambda)]^{-1}a=-\lambda P_E W[I-\lambda T^*]^{-1}a$$

and, according to (2.3), we have

$$\Theta_\Gamma(\lambda)=-W\Theta_T(\lambda)$$

which proves the lemma.

Our aim is to describe the elements of E and the form of the operator W in the Nagy-Foiaş functional model of T.

Let for, U, acting on K, be the minimal unitary dilation of T, $L=\overline{[U-T]H}$, $L_*=\overline{U[U^*-T^*]H}$ and ω,ω_* be the unitary operators from $\mathcal{D}_T$, $\mathcal{D}_{T^*}$ on L, L_* defined respectively, on the dense subspaces, by $\omega D_T h=[U-T]h$, $\omega_* D_{T^*}h=U[U^*-T^*]h$, $h\varepsilon H$.

LEMMA 3. *We have* $L=UE$, $L_*=W^*E$ *and* $\omega D_T h=UD_T h$, $\omega_* D_{T^*}h=D_{T^*}h$.

PROOF. As for $a\varepsilon E$ we have $Ta=0$, $T^*W^*a=0$ using Lemma 1 we obtain

$$\omega D_T h = \omega D_T^2 h = \omega D_T P_E h = [U-T]P_E h = UP_E h = UD_T h,$$
$$\omega_* D_{T^*} h = \omega_* D_{T^*}^2 h = U[U^*-T^*]D_{T^*}h = U[U^*-T^*]W^*P_E Wh = W^*P_E Wh = D_{T^*}h,$$

which proves the lemma.

Let us recall further some elements from the Nagy-Foias construction of the functional model for T (cf.[2], Ch.VI, Sec.2). The subspaces L and L_* are wandering subspaces for U and

$$K = M(L_*) \oplus R,$$
$$K_+ = M_+(L_*) \oplus R = H \oplus M_+(L),$$
$$P^{L_*} M_+(L) \subset M_+(L_*),$$
$$R = [I-P^{L_*}]K = [I-P^{L_*}]M(L),$$

where for a wandering subspace F for U, $M(F)$ stands for $\bigoplus_{-\infty}^{\infty} U^n F$, $M_+(F)$ for $\bigoplus_0^{\infty} U^n F$ and P^F for the orthogonal projection from K onto $M(F)$. For the bilateral shift U acting on $M(F)$, Φ_F denotes its Fourier representation, i.e. the unitary operator from $M(F)$ onto $L^2(F)$ defined by

$$\Phi_F\left(\sum_{-\infty}^{\infty} U^n f_n\right) = \sum_{-\infty}^{\infty} e^{int} f_n, \qquad f_n \in F, \quad \sum_{-\infty}^{\infty} \|f_n\|^2 < \infty.$$

Let Q be the contraction from $M(L)$ into $M(L_*)$ defined by $Q = P^{L_*}|M(L)$. Then there exists a contractive analytic function $\{L, L_*, \Theta_L(\lambda)\}$ such that

$$\Theta_L \Phi_L k = \Phi_{L_*} P^{L_*} k, \qquad k \in M(L). \tag{2.4}$$

We have

$$\Theta_T(\lambda) = \omega_*^{-1} \Theta_L(\lambda)\omega. \tag{2.5}$$

Further, if $\Delta_L(e^{it}) = [I-\Theta_L(e^{it})^*\Theta_L(e^{it})]^{1/2}$ the operator Φ_R defined on the dense subspace $[I-P^{L_*}]M(L)$ of R by

$$\Phi_R(I-P^{L_*})k = \Delta_L \Phi_L k, \qquad k \in M(L) \tag{2.6}$$

is unitary from R onto $\overline{\Delta_L L^2(L)}$ and $\Phi_R U|R = e^{it}\Phi_R$. Moreover, the operator Φ from K onto $L^2(L_*) \oplus \overline{\Delta_L L^2(L)}$ defined by

$$\Phi k = \Phi_{L_*} P^{L_*} k \oplus \Phi_R (I-P^{L_*})k, \qquad k \in K \tag{2.7}$$

is unitary and

$$\Phi H = [L_+^2(L_*) \oplus \overline{\Delta_L L^2(L)}] \ominus [\Theta_L f \oplus \Delta_L f,\ f \in L_+^2(L)]. \tag{2.8}$$

If $K' = \Phi K$, $H' = \Phi H$, $U' = \Phi U \Phi^*$, $T' = \Phi T \Phi^* | H'$ then

$$\begin{cases} U' u \oplus v = e^{it} u \oplus e^{it} v, & u \oplus v \in K' \\ T'^* u \oplus v = e^{-it}[u - u(0)] \oplus e^{-it} v, & u \oplus v \in H'. \end{cases} \tag{2.9}$$

Having in mind the form of L and L_* given by Lemma 3, we prove:

LEMMA 4. *For* $a\in E$ *we have*

$$\Phi a=e^{-it}\Theta_L Ua\oplus e^{-it}\Delta_L Ua,$$

$$\Phi W^*a=W^*a\oplus 0$$

$$\Theta_\Gamma(\lambda)a=-W\Theta_L(\lambda)Ua, \qquad \lambda\in D.$$

In the second side of the equalities, Ua *means the function in* $L^2_+(L)$ *identically equal to* $Ua\in L$ *and* W^*a *means the function in* $L^2_+(L_*)$ *identically equal to* $W^*a\in L_*$.

PROOF. For $a\in E$ we have $Ua\in L$ and according to (2.7)

$$\Phi Ua=\Phi_{L_*}P^{L_*}Ua\oplus\Phi_R[I-P^{L_*}]Ua=$$
$$=\Theta_L\Phi_L Ua\oplus\Delta_L\Phi_L Ua=\Theta_L Ua\oplus\Delta_L Ua.$$

Hence

$$\Phi a=U'^*U'\Phi a=U'^*\Phi Ua=U'^*\Theta_L Ua\oplus\Delta_L Ua=$$
$$=e^{-it}\Theta_L Ua\oplus e^{-it}\Delta_L Ua.$$

For $a\in E$ we have $W^*a\in L_*$ so

$$\Phi W^*a=\Phi_{L_*}P^{L_*}W^*a\oplus\Phi_R(I-P^{L_*})W^*a=W^*a\oplus 0.$$

According to (2.5) and Lemmas 1,2,3 we have

$$\Theta_\Gamma(\lambda)a=-W\Theta_T(\lambda)a=-W\omega_*^{-1}\Theta_L\omega a=-W\Theta_L(\lambda)Ua.$$

The lemma is proved.

Let us denote now by the same X the unitary operators from L_* onto E defined by $XW^*a=-a$, from L onto E defined by $XUa=a$, and their natural extensions (by intertwining with e^{it}) to unitary operators from $L^2(L_*)$ onto $L^2(E)$, from $L^2(L)$ onto $L^2(E)$ and from $L^2(L_*)\oplus L^2(L)$ onto $L^2(E)\oplus L^2(E)$. For every $f\in L^2(L)$, using Lemma 4, we obtain

$$(X\Theta_L f)(e^{it})=X\Theta_L(e^{it})f(e^{it})=-W\Theta_L(e^{it})f(e^{it})=$$
$$=WW^*\Theta_\Gamma(e^{it})U^*f(e^{it})=\Theta_\Gamma(e^{it})Xf(e^{it})$$

and

$$(X\Delta_L^2 f)(e^{it})=X\Delta_L^2(e^{it})f(e^{it})=$$
$$=X[I-\Theta_L(e^{it})^*\Theta_L(e^{it})]f(e^{it})=\Delta_\Gamma^2(e^{it})Xf(e^{it}).$$

Hence

$$X\Theta_L=\Theta_\Gamma X, \qquad X\Delta_L=\Delta_\Gamma X. \tag{2.10}$$

It results

$$XK'=L^2(E)\oplus\overline{\Delta_\Gamma L^2(E)}$$
$$XH'=[L^2_+(E)\oplus\overline{\Delta_\Gamma L^2(E)}]\ominus[\Theta_\Gamma f\oplus\Delta_\Gamma f,\ f\in L^2_+(E)].$$

Setting $\hat{K}=XK'$, $\hat{H}=XH'$, $\hat{U}=XU'X^*$, $\hat{T}=XT'X^*|\hat{H}$ we have

$$\hat{U}u\oplus v=e^{it}u\oplus e^{it}v, \qquad u\oplus v\in\hat{K} \tag{2.11}$$
$$\hat{T}^*u\oplus v=e^{-it}[u-u(0)]\oplus e^{-it}v, \qquad u\oplus v\in\hat{H}. \tag{2.12}$$

In addition for any $a\in E$ we have

$$X\Phi a=X[e^{-it}\Theta_L Ua\oplus e^{-it}\Delta_L Ua]=$$
$$=e^{-it}\Theta_\Gamma XUa\oplus e^{-it}\Delta_\Gamma XUa=e^{-it}\Theta_\Gamma a\oplus e^{-it}\Delta_\Gamma a,$$
$$X\Phi W^*a=X[W^*a\oplus 0]=-a\oplus 0.$$

We obtain

$$X\Phi a=e^{-it}\Theta_\Gamma a\oplus e^{-it}\Delta_\Gamma a,\quad X\Phi W^*a=-a\oplus 0, \qquad a\in E. \tag{2.13}$$

Let now V be the isometry form E into $\hat{H}$ defined by $V=X\Phi|E$. Then

$$Va=e^{-it}\Theta_\Gamma a\oplus e^{-it}\Delta_\Gamma a, \qquad a\in E \tag{2.14}$$

and for any $u\oplus v\in\hat{H}$ and $a\in E$ we have

$$(V^*u\oplus v,a)_E=(u\oplus v,Va)_{\hat{H}}=$$
$$=(u\oplus v,e^{-it}\Theta_\Gamma a\oplus e^{-it}\Delta_\Gamma a)=$$
$$=(e^{it}\Theta_\Gamma^* u+e^{it}\Delta_\Gamma v,a)_{L^2(E)}.$$

Hence

$$V^*u\oplus v=\frac{1}{2\pi}\int_0^{2\pi} e^{it}[\Theta_\Gamma(e^{it})^*u(e^{it})+\Delta_\Gamma(e^{it})v(e^{it})]dt,\ u\oplus v\in\hat{H}. \tag{2.15}$$

Let $\hat{W}$ be the unitary operator on $\hat{H}$ defined by $\hat{W}=X\Phi W\Phi^*X^*|\hat{H}$. Then for any $a\in E$ we have

$$(V^*\hat{W}^n Va,a)=(\hat{W}^n Va,Va)=(X\Phi W^n\Phi^*X^*X\Phi a,\ X\Phi a)=(W^n a,a)=(\Gamma(n)a,a).$$

Thus

$$\Gamma(n)=V^*W^nV, \qquad n\in\mathbb{Z}. \tag{2.16}$$

Since

$$\bigvee_{-\infty}^{\infty}\hat{W}^n VE=\bigvee_{-\infty}^{\infty}X\Phi W^n\Phi^*X^*X\Phi E=\bigvee_{-\infty}^{\infty}X\Phi W^n E=X\Phi\bigvee_{-\infty}^{\infty}W^n E=X\Phi H=\hat{H}$$

we obtain

$$\hat{H}=\bigvee_{-\infty}^{\infty}\hat{W}^n VE. \tag{2.17}$$

Also

$$\hat{W}^*[I-VV^*]=X\Phi W^*\Phi^*X^*[I-X\Phi V^*]=X\Phi W^*[I-P_E]\Phi^*X^*=X\Phi T\Phi^*X^*,$$

so

$$\hat{T}=W^*[I-VV^*]. \tag{2.18}$$

For any $u\oplus v\in\hat{H}$ and $a\in E$, using (2.13) we obtain

$$(V^*\hat{W}u\oplus v,a)_E=(\hat{W}u\oplus v,Va)_{\hat{H}}=(Wu\oplus v,\ X\Phi a)_{\hat{H}}=$$
$$=(W\Phi^*X^*u\oplus v,a)_H=(u\oplus v,\ X\Phi W^*a)_{\hat{H}}=(u\oplus v,-a+0)_{\hat{H}}=$$
$$=-(u,a)_{L^2(E)}=-(u(0),a)_E.$$

It results

$$(2.19)\qquad V^*Wu\oplus v=-u(0),\qquad u\oplus v\in\hat{H}.$$

From (2.12), (2.18) and (2.19) we obtain

$$Wu\oplus v=\hat{T}^*u\oplus v+VV^*\hat{W}u\oplus v=T^*u\oplus v-Vu(0)=$$
$$=e^{-it}[u-u(0)]\oplus e^{-it}v-e^{-it}\Theta_\Gamma u(0)\oplus e^{-it}\Delta_\Gamma u(0)=$$
$$=e^{-it}[u-\Theta_\Gamma u(0)-u(0)]\oplus e^{-it}[v-\Delta_\Gamma u(0)].$$

The proof of the theorem is complete.

REFERENCES

1. Brodskiĭ, M.S.: Unitary operator nodes and their characteristic function (Russian), *Uspekhi Matem.Nauk* 33:4(1978), 141-168.

2. Sz.-Nagy, B; Foiaş, C.: *Harmonic analysis of operators on Hilbert space*, Acad.Kiadó-Budapest, North Holland Company, Amsterdam, London, 1970.

Ion Suciu
Department of Mathematics,
INCREST,
Bdul Păcii 220, 79622 Bucharest,
Romania.

THE LEVINSON ALGORITHM IN LINEAR PREDICTION

Dan Timotin

The prediction theory of infinite-variate processes, as presented in [3], is a natural generalization of the classical prediction theory of Wiener and Masani. One of the main problems in this theory is the actual computation of the "predictible part" of the process. A recurrent algorithm, due to Levinson ([2]) in the univariate case, can be used for this purpose. The present note extends the algorithm to the most general frame, by using its geometrical structure. For other extensions and comments on the Levinson algorithm, also in relation with solving Toeplitz systems of equations, see [1].

1. We recall the basic definitions of prediction theory in complete correlated actions ([3], [4]). Suppose E is a separable Hilbert space, H is a right $L(E)$-module and $\Gamma: H\times H\to L(E)$ satisfies the following axioms:

(i) $\Gamma[h,h]\geq 0$; $\Gamma[h,h]=0 \Rightarrow h=0$

(ii) $\Gamma[h_1,h_2]^*=\Gamma[h_2,h_1]$

(iii) $\Gamma[\sum_i h_i A_i, \sum_j g_j B_j]=\sum_{i,j} A_i^*\Gamma[h_i,g_j]B_j$.

Such a triple $\{E,H,\Gamma\}$ is called a *correlated action* of $L(E)$ on H. In [3] it is also shown that every correlated action can be embedded into a "typical" one; that is, there exists a Hilbert space K and an algebraic embedding $h\to V_h$ of the right $L(E)$-module H into the right $L(E)$-module $L(E,K)$ such that $\Gamma[h_1,h_2]=V_{h_1}^* V_{h_2}$ $(h_1,h_2\in H)$. Moreover, K is uniquely determined (up to a unitary equivalence) by requiring that $K=\bigvee_{\substack{h\in H\\ a\in E}} V_h a$; it is called the *measuring space* of the correlated action. A correlated action is *complete* if the image of H under this embedding is the whole of $L(E,K)$. In a complete correlated action one can define ([4]) the "projection" of an element $h \in H$ onto a submodule $H_1 \subset H$; $P_{H_1} h$ is the unique element of H such that

$$\Gamma[h-P_{H_1}h, h_1]=0 \qquad \text{for any } h_1 \in H_1$$

and

$$\Gamma[h-P_{H_1}h, h-P_{H_1}h] = \inf_{h_1 \in H_1} \Gamma[h-h_1, h-h_1]$$

where the infimum is taken in the ordered set of positive operators on E. It is defined by the relation $V_{P_{H_1}h} = P_{K_1} V_h$, where $K_1 = \bigvee_{g \in H_1} V_g E$. Note that $P_{H_1} h$ may in fact not belong to H_1; but, if H_1' is the submodule of H generated by H_1 and $P_{H_1} h$, then $P_{H_1'} = P_{H_1}$.

A stationary process in a correlated action is a sequence $\{f_n\} \subset H$ such that $\Gamma[f_n, f_m]$ depends only on the difference $n-m$. Two stationary processes $\{f_n\}, \{g_n\}$ are *cross-correlated* if $\Gamma[f_n, g_m]$ depends only on $n-m$.

We shall consider further a complete correlated action $\{E, H, \Gamma\}$. Suppose we have a stationary process $\{f_n\} \subset H$. Denote

$$H_f^n = \{h \in H,\ h = \sum_{k \geq 0} f_{n-k} A_k, \text{ where } f_k \neq 0 \text{ only for a finite number of values of } k\}.$$

Put $\hat{f}_{n+1} = P_{H_f^n} f_{n+1}$, and $g_{n+1} = f_{n+1} - \hat{f}_{n+1}$. Then $\hat{f}_{n+1}$ is called the "predictible part" of f_{n+1}, and g_{n+1} the "innovation part". One of the problems of prediction is that of determining $\hat{f}_{n+1}$, which may be called the "best estimation" of f_{n+1} by elements in the submodule H_f^n. More precisely, we have to find a sequence $(A_0^N, A_1^N, \ldots\ldots\ldots\ldots, A_{m(N)}^N)$ of operators in $L(E)$, such that, if $h_n^N = \sum_{k=0}^{m(N)} f_{n-k} A_k^N$, then $V_{h_n^N} \underset{N \to \infty}{\to} V_{\hat{f}_{n+1}}$ strongly in $L(E,K)$. (This convergence may also be characterized without using the measuring space K; indeed $(V_{h_n^N} - V_{\hat{f}_{n+1}})a \to 0$ for every $a \in E$ is easily seen to be equivalent to $\Gamma(V_{h_n^N} - V_{\hat{f}_{n+1}}, V_{h_n^N} - V_{\hat{f}_{n+1}})$ tends strongly (or weakly) to 0 in $L(E)$.) In the next section we present the general algorithm for obtaining this sequence. A supplementary result for a particular case will be obtained in the last section.

2. Denote by

$$\mathcal{H}_f^n(N)=\{h\in\mathcal{H},\ h=\sum_{k=0}^{N} f_{n-k}A_k\}.$$

The Levinson algorithm is a method of obtaining successively the projection $\hat{f}_1(N)$ of f_1 onto the finitely generated submodules $\mathcal{H}_f^1(N)$. The coefficients should be obtained in terms of the correlation sequence $\Gamma[f_n,f_m]=\Gamma(n-m)$. Note that, since the process is stationary, analogous formulae would give the projection of any f_{n+1} onto $\mathcal{H}_f^n(N)$.

The algorithm will use simultanously the reverse stationary process $\{f_n'\}$, defined by $f_n'=f_{-n}$. We shall denote all the corresponding objects by primes; we have also

$$\Gamma[f_n',f_{n+k}']=\Gamma[f_{-n},f_{-n-k}]=$$
$$=\Gamma[f_{-n-k},f_{-n}]^*=\Gamma[f_n,f_{n+k}]^*.$$

For S a positive operator on $L(E)$, we shall denote by S^{-1} the (possibly unbounded) operator $\phi(S)$, where $\Phi(t)=\frac{1}{t}$ for $t>0$, and $\phi(t)=0$ for $t\le 0$. Now, if $h\in\mathcal{H}$, $A\in L(E)$, then $hS^{-1}A$ does not make sense as it stands,but we shall use this notation in the sequel only when we are sure that $V_hS^{-1}A\in L(E,K)$ (is actually bounded). (Note that in this case $V_hS^{-1}A$ can always be obtained as a strong limit of uniformly bounded operators of the form V_hS_nA, $S_n\in L(E,K)$.) In fact, the notation will appear as a consequence of the following lemma, whose statement is well-known in case dim $E=1$ (the Hilbert space case):

LEMMA. *If* $g,h\in\mathcal{H}$, $\mathcal{H}_g$ *is the submodule of* $\mathcal{H}$ *generated by* g, *then*

$$P_{\mathcal{H}_g}h=g\Gamma[g,g]^{-1}\Gamma[h,g].$$

The proof follows simply from the fact that, if $K_g=\overline{V_gE}$, then

$$P_{K_g}=V_g(V_g^*V_g)^{-1}V_g^*.$$

The main result of this paper is

THEOREM. *We have*

$$\hat{f}_{n+1}(N)=\sum_{k=0}^{N} f_{n-k}A_k^N$$

$$\hat{f}'_{n+1}(N) = \sum_{k=0}^{N} f'_{n-k} A'^{N}_{k} ,$$

the coefficients $A^N_k (0 \le k \le N)$ *being obtained recurrently as follows:*

$$A^0_0 = \Gamma(0)^{-1}\Gamma(1)$$

$$A'^{0}_{0} = \Gamma(0)^{-1}\Gamma(1)^*$$

and, if

$$E_N = \Gamma[f_{-N-1} - \sum_{k=0}^{N} f_{-N+k} A'^{N}_{k}, f_{-N-1} - \sum_{k=0}^{N} f_{-N+k} A'^{N}_{k}]^{-1} . \Gamma[f_1, f_{-N-1} - \sum_{k=0}^{N} f_{-N+k} A'^{N}_{k}],$$

$$E'_N = \Gamma[f_{N+1} - \sum_{k=0}^{N} f_{N-k} A^{N}_{k}, f_{N+1} - \sum_{k=0}^{N} f_{N-k} A^{N}_{k}]^{-1} . \Gamma[f_{-1}, f_{N+1} - \sum_{k=0}^{N} f_{N-k} A^{N}_{k}],$$

then

$$A^{N+1}_{N+1} = E_N , \quad A'^{N+1}_{N+1} = E'_N$$

$$A^{N+1}_{k} = A^{N}_{k} - A'^{N}_{N-k} E_N \qquad \text{for } 0 \le k \le N$$

$$A'^{N+1}_{k} = A'^{N}_{k} - A^{N}_{N-k} E'_N \qquad \text{for } 0 \le k \le N.$$

PROOF. The formula for N=0 is a direct consequence of the lemma (applied to f_{n+1} and f_n instead of h and g). To obtain the general formulae, note that $H^n_f(N+1)$ is generated by $H^n_f(N)$ and f_{-N-1}, and the projection of $f_{-N-1} = f'_{N+1}$ on $H^n_f(N)$ is exactly $\hat{f}'_{N+1}(N)$. By the remarks in Section 1, the projection onto the submodule generated by $H^n_f(N)$ and $\hat{f}'_{N+1}(N)$ is equal to $P_{H^n_f(N)}$. If $g = f_{-N-1} - \hat{f}'_{N+1}(N)$, then $H^n_f(N)$ and H_g (the submodule generated by g) are Γ- orthogonal, whence it follows easily that

$$P_{H^n_f(N+1)} = P_{H^n_f(N)} + P_{H_g}.$$

We apply now the lemma in order to compute P_{H_g}, and the desired formulae follow.

REMARKS.1. It is obvious that E_N, E'_N can be computed in terms of the coefficients A_k^N, $A_k'^N$ and the correlation function $\Gamma(n)$.

2. If $\dim E<\infty$, then S^{-1}, as defined above, is bounded for any positive operator S, so all coefficients A_k^N belong to $L(E)$.

3. If $\dim E=1$, we do not need the reverse process, since we obtain easily $A_k'^N=\overline{A_k^N}$.

COROLLARY. *The predictible part* $\hat{f}_{n+1}$ *can be obtained as*

$$\hat{f}_{n+1}=\lim_{N\to\infty} \hat{f}_{n+1}(N)$$

where the limit means strong limit in $L(E,K)$.

The proof is immediate, noting that

$$P_{H_f^n}=\text{s-}\lim_{N\to\infty} P_{H_f^n(N)}.$$

We remark that the "error operator"

$$\Delta_f=\Gamma[f_1-\hat{f}_1,\ f_1-\hat{f}_1]^{1/2}$$

may also be obtained as

$$\Delta_f=\text{s-}\lim_{N\to\infty} \Gamma[f_1-\hat{f}_1(N),f_1-\hat{f}_1(N)]^{1/2}.$$

3. In [3] it is proved that, if the spectral distribution F of the process (the semi-spectral measure which has the $\Gamma(k)$ as Fourier coefficients) satisfies the relation

$$c\,dt\le dF\le\frac{1}{c}dt \qquad \text{for some } c>0$$

then there exists a bounded , invertible outer function $\{E,E,\Theta(\lambda)\}$, such that the process is unitary equivalent to the following functional model:

$$H=L(E,L^2(E)),$$
$$f_n(a)(e^{it})=e^{-int}\Theta(e^{it})a,$$
$$g_n(a)(e^{it})=e^{-int}\Theta(0)a.$$

If we denote by $\{E,E,\Omega(\lambda)\}$ the operator analytic function that satisfies $\Omega(\lambda)\Theta(\lambda)=\Theta(\lambda)\Omega(\lambda)=I_E$, and $\Omega(\lambda)=\sum_{k=0}^{\infty}\lambda^k\Omega_k$, then we have ([3]) the formula

$$g_n=\sum_{k=0}^{\infty} f_{n-k}\,\Omega_k\,\Theta(0)$$

where the series is convergent in the strong topology of $L(E,L^2(E))$. The formula is a consequence of the obvious relation

$$g_n = e^{-int}\Theta(0) = \Theta(e^{it})\Omega(e^{it})\Theta(0).$$

Now, if we denote

$$g_n(N) = f_n - \hat{f}_n(N)$$

the theorem in Section 2 gives

$$g_{n+1}(N) = f_{n+1} - \sum_{k=0}^{N} f_{n-k}A_k^N$$

which may be written, if $\Omega_N(\lambda) = (I_E - \sum_{k=1}^{N+1} \lambda^k A_{k-1}^N)\Theta(0)^{-1}$ as

$$g_n(N) = \Theta(e^{it})\Omega_N(e^{it})\Theta(0).$$

Since $g_n(N) \to g_n$ for $N \to \infty$ (strongly in $L(E,L^2(E))$, it follows that $\Omega_N(e^{it})a \to \Omega(e^{it})a$, in $L^2(E)$ for any $a \in E$. We know that Ω is an outer function.

A complementary result is the following

PROPOSITION. Ω_N *is an outer function for any* N.

PROOF. For $a,b \in E$, $k=1,\dots,N+1$ the definition of $g_n(N)$ implies that $\Theta(e^{it})\Omega_N(e^{it})a$ is orthogonal (in $L^2(E)$) to $e^{ikt}\Theta(e^{it})b$. So

$$\frac{1}{2\pi}\int_{-\pi}^{\pi} (\Theta(e^{it})^* \Theta(e^{it})\Omega_N(e^{it})a, e^{ikt}b)\,dt = 0,$$

for $k=1,\dots,N+1$.

$\Theta(e^{it})^* \Theta(e^{it})\Omega_N(e^{it})$ is an essentially bounded measurable function with values in $L(E)$ whose Fourier coefficients vanish for $k=1,\dots,N+1$; so there exist bounded analytic functions $\Phi_1(\lambda)$, $\Phi_2(\lambda)$, such that

$$\Theta(e^{it})^* \Theta(e^{it})\Omega_N(e^{it}) = \Phi_1(e^{it})^* + e^{i(N+2)t}\Phi_2(e^{it}).$$

Since Ω_N is a polynomial of degree at most N, we may find an analytic function $\Phi_3(\lambda)$, such that

$$\Omega_N(e^{it})^* \Theta(e^{it})^* \Theta(e^{it})\Omega_N(e^{it}) = \Omega_N(e^{it})^*\Phi_1(e^{it})^* + e^{it}\Phi_3(e^{it}).$$

The left hand side has values self adjoint operators, while the right hand side is the sum of an analytic and a co-analytic function. By comparing Fourier coefficients of both sides, we obtain

$$\Omega_N(e^{it})^* \ \Theta(e^{it})^*\Theta(e^{it})\Omega_N=$$
$$=2\mathrm{Re}\Omega_N(e^{it})\Phi_1(e^{it})-\Omega_N(0)^* \ \Theta(0)^*\Theta(0)\Omega_N(0).$$

But $\Omega_N(0)=I_E$, $\Theta(0)\geq 0$, so

$$2\mathrm{Re}\Omega_N(e^{it})\Phi_1(e^{it})=$$
$$=\Omega_N(e^{it})^* \ \Theta(e^{it})^* \ \Theta(e^{it})\Omega_N(e^{it})+\Theta(0)^*\Theta(0)\geq$$
$$\geq \Theta(0)^*\Theta(0).$$

Since $\Theta(\lambda)$ has a bounded inverse, $\Theta(0)^*\Theta(0)$ is bounded below, and so is the left hand side of the inequality. This property extends by the maximum principle also for $\mathrm{Re}\Omega_N(\lambda)\Phi_1(\lambda)$, $|\lambda|<1$. Therefore $\Omega_N(\lambda)\Phi_1(\lambda)$ has a bounded inverse on all $|\lambda|<1$; so does then $\Omega_N(\lambda)$, which is consequently an outer function.

REMARK. The case $\dim E=1$ of this proposition follows from a classical result of Szegö (see [1]).

REFERENCES

1. Kailath, T.: A view of three decades of linear filtering theory, *IEEE Trans.Information Theory* IT-20(1974), 146-181.

2. Levinson, N.: The Wiener rms (root-mean-square) error criterion in filter design and prediction, *J.Math.Phys.* 25(1946), 261-278.

3. Suciu, I.; Valuşescu, I.: Factorization theorems and prediction theory, *Rev.Roumaine Math.Pures Appl.*23(1978), 1393-1423.

4. Suciu, I.; Valuşescu, I.: A linear filtering problem in complete correlated actions, *J.Multivariate Analysis* 9(1979), 599-613.

5. Wiener, N.; Masani, P.: The prediction theory of multivariate stochastic processes. I , *Acta Math.*98(1957), 111-150.

6. Wiener, N.; Masani, P.: The prediction theory of multivariate stochastic processes. II, *Acta Math.*99(1958), 93-139.

Dan Timotin
Department of Mathematics,
INCREST,
Bdul Păcii 220, 79622 Bucharest,
Romania.

GEOMETRIC INTERPRETATION OF THE ESSENTIAL MINIMUM MODULUS

Jaroslav Zemánek

The present note is devoted to a geometric characteristic, the so called essential minimum modulus. We point out here some analogies with the previous results which became evident after the appearance of R.Bouldin's paper [2].

Let X be a complex Banach space. We denote by $\mathcal{B}(X)$ the Banach algebra of bounded linear operators on X. Given an operator T in $\mathcal{B}(X)$ we put

$$m(T)=\inf\{||Tx||:||x||=1\} .$$

This quantity is usually called the minimum modulus of T. It is well known that $m(T)>0$ if and only if the nullspace of T is zero and the range of T is closed. Let $\mathcal{M}(X)$ denote the set of operators T with $m(T)>0$. It is clear that the set $\mathcal{M}(X)$ is open in $\mathcal{B}(X)$. More precisely, we have shown in [4] that the value of $m(T)$ is exactly the distance from the operator T to the closed set $\mathcal{B}(X)\setminus\mathcal{M}(X)$. Further, denoting by $b(T)$ the supremum of all $\varepsilon\geq 0$ such that $m(T-\lambda I)>0$ for $|\lambda|<\varepsilon$, we have obtained in [4] the following formula

$$b(T)=\lim_{n\to\infty} m(T^n)^{1/n} . \tag{1}$$

If X is a Hilbert space then the quantity $m(T)$ admits another interpretation. Namely, it is the minimal point in the spectrum of the operator $(T^*T)^{1/2}$. R.Bouldin [2] has recently studied the essential minimum modulus $m_e(T)$ which is defined as the minimal point in the essential spectrum of $(T^*T)^{1/2}$. In this note we show a geometric interpretation of this quantity analogous to that of $m(T)$. This observation makes it possible to define the essential minimum modulus for any Banach space operator and suggests a number of conjectures analogous to the known results.

Let $M_+(X)$ denote the set of all operators which have finite dimensional nullspace and closed range. It is well known that the set $M_+(X)$ is open in $B(X)$ and clearly contains $M(X)$. In the literature the operators in $M_+(X)$ are usually called Φ_+-operators.

THEOREM. *Let* T *be a bounded operator on a Hilbert space* H. *Then the essential minimum modulus* $m_e(T)$ *is exactly the distance from the operator* T *to the closed set* $B(H)\setminus M_+(H)$.

PROOF. First we have to show that if $||T-S||<m_e(T)$ then $m_e(S)>0$. To this end we consider the images of T and S in the Calkin algebra. Since the latter is a C^*-algebra, its elements can be represented as operators on some Hilbert space K having some norm $|\cdot|$. Since $|T-S|=||T-S||_e\leq||T-S||$ we conclude that $m_e(S)>0$ as follows. By the above mentioned spectral characterization of the minimum modulus in the Hilbert space situation we have $m_e(S)=\inf\{|Sx|:x\in K, |x|=1\}$. But here $|Sx|\geq|Tx|-|(T-S)x|\geq$ $\geq m_e(T)-|T-S|>0$ and so we are done.

Secondly, given $\varepsilon>0$ we have to find an operator S in $B(H)\setminus$ $\setminus M_+(H)$ such that $||T-S||\leq m_e(T)+\varepsilon$. To this end we consider the spectral resolution $E(\cdot)$ of the operator $(T^*T)^{1/2}$. Then the subspace $K=E[\ 0\ , m_e(T)+\varepsilon)H$ has infinite dimension, cf. [2], Theorem 2(ii). Now let the operator S be defined by $Sy=0$ for y in K, and $Sz=Tz$ for z in the orthogonal complement of K. Hence the operator S is in $B(H)\setminus M_+(H)$. Let $x=y+z$ be an arbitrary vector of norm one in H, with y and z as above. Then $(T-S)x=Ty$ hence $||(T-S)x||=||Ty||\leq m_e(T)+\varepsilon$. The last inequality is a consequence of Theorem 75.2 on p.253 in [1]. The proof is complete.■

The preceding theorem and the previously mentioned geometric interpretation of the minimum modulus suggest the following definition of the essential minimum modulus for any Banach space operator. It is natural to put

$$m_e(T)=\mathrm{dist}(T,B(X)\setminus M_+(X)) \ . \tag{2}$$

Now let $b_e(T)$ be the supremum of all $\varepsilon \geq 0$ such that $m_e(T-\lambda I)>0$ for $|\lambda|<\varepsilon$. Is it possible to express the radius $b_e(T)$ in terms of the essential minimum modulus $m_e(T)$ by a formula analogous to (1)?

Also similar questions arise concerning the dual problem of surjectivity and essential surjectivity. Let $Q(X)$ denote the set of all operators which map X onto X, and let $Q_-(X)$, a larger set, consist of all operators on X having the range of finite codimension (the so called Φ_--operators). We put $q(T)=\mathrm{dist}(T,B(X)\setminus Q(X))$ and $q_e(T)=\mathrm{dist}(T,B(X)\setminus Q_-(X))$.

If X is a Hilbert space it is easy to see, using the above theorem, that $q_e(T)=m_e(T^*)$. In this case R.Bouldin [2] has expressed the distance from an operator T to the set of invertible (or Fredholm) operators in terms of the quantities $m_e(T)$ and $m_e(T^*)$. Is it possible to extend his results, formulated in terms of $m_e(T)$ and $q_e(T)$, to the general situation when X is a Banach space?

REFERENCES

1. Akhiezer, N.I.; Glazman, I.M.: *Theory of linear operators on Hilbert space* (Russian), Nauka, Moskva, 1966.
2. Bouldin, R.: The essential minimum modulus, *Indiana Univ. Math.J.* 30(1981), 513-517.
3. Makai, E.; Zemánek, J.: Geometrical means of eigenvalues, *J.Operator Theory*, 7(1982), 173-178.
4. Makai, E.; Zemánek, J.: The surjectivity radius, packing numbers and boundedness below of linear operators, *Integral Equations Operator Theory*, submitted.
5. Zemánek, J.: Generalisations of the spectral radius formula, *Proc.Roy.Irish Acad.* 81A(1981), 29-35.
6. Zemánek, J.: The essential spectral radius and the Riesz part of spectrum, in *Functions, Series, Operators* (Proc. Internat.Conf., Budapest, 1980), Colloq.Math.Soc.János Bolyai, to appear.

Jaroslav Zemánek
ul.Nerudova 14,
274-01 Slaný,
Czechoslovakia.

OPERATOR THEORY: ADVANCES AND APPLICATIONS

Published Volumes

OT 1:
H. Bart, I. Gohberg, and M.A. Kaashoek
Minimal Factorization of Matrix and Operator Functions
1979. 236 pages. Paperback
ISBN 3-7643-1139-8

OT 2:
Topics in Modern Operator Theory
5th International Conference on Operator Theory, Timisoara and Herculane (Romania), June 2-12, 1980
Edited by C. Apostol, R.G. Douglas, B. Sz.-Nagy, D. Voiculescu
Managing Editor: Gr. Arsene
1981. 336 pages. Hardcover
ISBN 3-7643-1244-0

OT 3:
K. Clancey and I. Gohberg
Factorization of Matrix Functions and Singular Integral Operators
1981. 234 pages. Hardcover
ISBN 3-7643-1297-1

OT 4:
Toeplitz Centennial
Toeplitz Memorial Conference in Operator Theory, Dedicated to the 100th Anniversary of the Birth of Otto Toeplitz
Tel Aviv, May 11-15, 1981
Edited by I. Gohberg
1982. 590 pages. Hardcover
ISBN 3-7643-1333-1

The series "Operator Theory: Advances and Applications" is devoted to the publication of current research in Operator Theory, including the full range of its applications.
Particular attention will be given to applications to classical analysis and the theory of integral equations as well as to numerical analysis, mathematical physics and mathematical methods in electrical engineering.

The book series supplements the journal "Integral Equations and Operator Theory" and has the same editorial board.